Game Theory Explained

A Mathematical Introduction with Optimization

Game Theory Explained

A Mathematical Introduction with Optimization

Christopher Griffin

Pennsylvania State University, USA

World Scientific

NEW JERSEY · LONDON · SINGAPORE · GENEVA · BEIJING · SHANGHAI · TAIPEI · CHENNAI

Published by

World Scientific Publishing Co. Pte. Ltd.

5 Toh Tuck Link, Singapore 596224

USA office: 27 Warren Street, Suite 401-402, Hackensack, NJ 07601

UK office: 57 Shelton Street, Covent Garden, London WC2H 9HE

Library of Congress Cataloging-in-Publication Data
Names: Griffin, Christopher, 1979- author.
Title: Game theory explained : a mathematical introduction with optimization /
    Christopher Griffin, Pennsylvania State University, USA.
Description: New Jersey : World Scientific, [2025] | Includes bibliographical references and index.
Identifiers: LCCN 2024035441 | ISBN 9789811297212 (hardcover) |
    ISBN 9789819812875 (paperback) | ISBN 9789811297229 (ebook for institutions) |
    ISBN 9789811297236 (ebook for individuals)
Subjects: LCSH: Game theory. | Mathematical optimization.
Classification: LCC QA269 .G75 2025 | DDC 519.3--dc23/eng/20241226
LC record available at https://lccn.loc.gov/2024035441

British Library Cataloguing-in-Publication Data
A catalogue record for this book is available from the British Library.

For any available supplementary material, please visit
https://www.worldscientific.com/worldscibooks/10.1142/13962#t=suppl

Desk Editors: Nambirajan Karuppiah/Rosie Williamson

Typeset by Stallion Press
Email: enquiries@stallionpress.com

This book is dedicated to Beverly Leech (Kate Monday), Joe Howard (George Frankly), and Toni DiBuono (Pat Tuesday) of *Mathnet* and every person who made *Square One TV* possible when I was a kid. Without you, this book would not exist.

Preface

Why did I write this book? This book started in 2010 as a series of lecture notes called "Game Theory: Penn State Math 486 Lecture Notes." Game theory was the second course I ever taught, and I wrote the notes during a cold State College winter. People always ask me, "Why do you write such detailed lecture notes?" And I always tell them, "Sheer terror!" As a young faculty member, one is always worried about something, and I figured if I had a well-prepared set of typed lecture notes, the chances of me making a bone-headed mistake in class would be minimized. When I started the class, I had intended to use Luce and Raiffa's book [1], with elements of Morris' book [2] added for good measure. But the further I got into the class, the more I realized I wanted to cover the material in a way that emphasized the connection to other parts of mathematics, especially optimization. Nisan, Roughgarden, Tardos, and Vazirani had published their book, *Algorithmic Game Theory* [3], a few years earlier, and there was a sudden emphasis on finding equilibria in games algorithmically. I was aware of the work from the 1960s on using optimization methods to find Nash equilibria; consequently, a rather unique course was born that emphasized both classical results from game theory and interesting results from optimization theory. The material was aimed squarely at undergraduates and (in my opinion) provided a deep but not impenetrable introduction to the mathematics of games that is often missing at the undergraduate level.

When I taught at the United States Naval Academy for a few years, I developed a course on "Computational Game Theory," where I continued to polish and refine the lecture notes. Finally, during the COVID-19 pandemic, a very nice editor named Rochelle from World Scientific contacted me and asked if I'd be willing to turn some of my notes into a book. I started with my notes on graph theory, and the result was my first book, *Applied Graph Theory*. With that project finished, I was finally ready to turn my notes on game theory into a book, and the result is this text. Between you and me, I'm glad I wrote this one second. The graph theory notes were more mature and easier to convert; I had found my voice (for the most part). Writing this book has been a bit like taking a trip down memory lane to a time when I was less confident in myself as a teacher and had not yet found my voice as a writer. Frankly, parts of the process were painful, but I'm reasonably pleased with the outcome. I will give you a fair warning: There are parts of this book that are dense. I don't like the "fluffy" approach to game theory, and I feel that such an approach hides too much of the beautiful intricacies of the subject and its deeper connections to the mathematical world. That being said, I've included plenty of examples and worked problems so that the density is broken up by practical applications of game theory, along with some mathematical interludes that I hope you enjoy.

How could you use this book? This book can be used entirely for self-study or in a classroom. I've had e-mails from all around the world saying that people have used the lecture note form for self-study, so it is possible. The book is really designed for a one-semester course and is geared toward undergraduates who have familiarity with differential vector calculus and matrices. The book emphasizes the proofs. However, many of these can be skipped on a first reading, though you will lose something if you skip too many of them. Almost all the proofs are derivations, though a few are by contradiction. So, while a "proofs class" is not needed, it won't hurt.

The book is organized into three parts and written in a "theorem-proof-example" style. There are remarks throughout and chapter notes that try to emphasize some history or other examples. Part 1 covers classical game theory, including games against the house (casino games) and utility theory. Part 2 covers the relationship between game theory and optimization. Part 3 covers cooperative game theory. There is an appendix that introduces evolutionary game theory by way of replicator dynamics. I used this material just after the pandemic, when we were running a week ahead of schedule in the course. In general, I favor a separate class on evolutionary games rather than wedging it into a course on game theory. Of course, part of my research focuses on evolutionary game theory, so I would say that. Each part emphasizes not only the results but also the proofs along with examples. A nice feature of the book is the relationship between Nash bargaining theory and multi-criteria optimization, which (as far as I know) is not covered in any other book.

Sample curriculum paths are shown in the following figure.

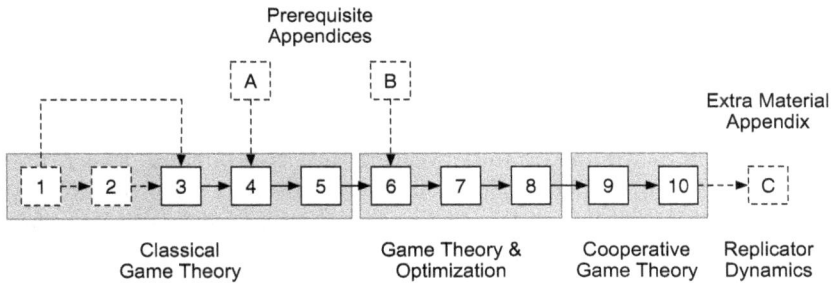

Classic route: The first time I taught this course, I went from Chapter 1 to Chapter 10, with limited information coming from the appendices. This is by far the easiest way to teach the course and gives a thorough understanding of classical game theory and its relationship to optimization. The material on optimization and probability is self-contained, so there is no need to use an external resource, unless so desired. It has been my experience that students like the material on games against the house in Chapter 1, which helps offset the formal introduction to probability that it provides.

Skip utility theory: When I taught game theory in 2021, I decided to skip Chapter 2. I like the material in it, but the students often find it a little dry. It is perfectly fine to skip Chapter 2 and go from Chapter 1 to Chapter 3. In this case, one could proceed at a slower pace and really emphasize all the proofs in the book or include Appendix C on evolutionary game theory, which is self-contained enough that students don't need to have seen differential equations (though it would help). In 2021, we covered the material in Appendix C and did most of the proofs.

Skip Chapters 1, 2 and (parts of) 3:   I've never done this because the students like the games in Chapter 1, but it is possible to skip Chapters 1 and 2, assuming that the students are familiar enough with probability theory. In that case, Chapter 3 could be condensed to just the material on game trees that do not include games of chance. (Essentially, you're skipping poker.) It's also possible to skip all of Chapter 3 and start immediately at Chapter 4 with normal form games. I would only do this if I wanted to really emphasize the proofs or review the material on calculus and matrices thoroughly. You would probably have to include Appendix C to make a 15-week course, though this could work in a 10-week term.

How do you handle optimization? The book is sensitive to the fact that optimization can be a niche interest in some math departments. Consequently, all the material on optimization is self-contained, including the coverage of the Karush–Kuhn–Tucker conditions. Penn State has no math course that covers these as a prerequisite for the game theory class, and students had no problems with them. Appendices A and B can also be used to help bring students up to speed on elements of matrix arithmetic and calculus that they may have forgotten or missed.

My favorite aspect of the book is the connection between games and optimization, especially the connection between optimization and cooperative games, which, I think, is wholly underemphasized. The quadratic programming method for finding Nash equilibria in general-sum bimatrix games is not the most "modern" way to solve such problems, but it is both understandable to undergraduates and can be performed on a computer algebra system. Moreover, it provides critical mathematical foundations for more advanced study in game theory, using, for example, the book by González-Díaz, García-Jurado, and Fiestras-Janeiro [4].

What isn't in this book?   Since this is geared toward a one-semester class, there are several omissions. There is no mention of dynamic or differential games, unless you consider multi-layered game trees to be a form of dynamic game, which I do not. Likewise, iterated games are not covered because they could form an entire text on their own. Cooperative game theory of the kind studied deeply in economics is only introduced in Chapter 10. It is not a feature of the book. Evolutionary game theory is covered in Appendix C.

As this is one of my main areas of research, I would prefer a separate treatment of the subject. However, the material on the replicator is both exciting and produces nice pictures, so it is worth putting into an appendix as "bonus material" to be used as needed. Formal treatment of evolutionarily stable strategies (ESS) is explicitly omitted. ESS are subtle, and in my experience, the students hate it. When presented poorly, it is literally a way to turn students off from evolutionary games.

Explicit optimization algorithms such as the simplex algorithm are not presented. Instead, I show how to solve optimization problems with a computer. For those readers interested in understanding the methods used by the computer for solving the optimization problems that arise in this book, there are ample references.

Acknowledgements

This book was created with LaTeX2e using Overleaf, TeXShop, BibDesk, and the grammar checker LanguageTool. Figures were created using Mathematica™ and Omnigraffle™, unless otherwise noted. In acknowledging those people who helped make this work possible, first let me say thank you to all the other scholars who have worked in game theory. Without your pioneering work, there would be nothing to write about. Also, I must thank individuals who found typos in my original lecture notes and wrote to tell me about them: James Fan, George Kesidis, Nicolas Aumar, Arlan Stutler, and Sarthak Shah. I would also like to thank Volmir Eugênio Wilhelm, who selflessly translated a version of my game theory lecture notes into Spanish. If there is anyone else I have forgotten, I apologize, but your contributions are appreciated. I would very much like to thank Andrew Belmonte, at Penn State, for encouraging me to turn my notes into a book, even though I ignored him for 14 years. Finally, I owe a debt to Rochelle Kronzek Miller of World Scientific, who gave me the final push to write this, and to my desk editor, Rosie Williamson, and her editorial staff, who dealt with the mechanics of getting this published. Thank you all.

About the Author

Christopher Griffin is a Research Professor at the Applied Research Laboratory (ARL), where he holds a courtesy appointment as Professor of Mathematics at Penn State. He was a Eugene Wigner Fellow in the Computational Science and Engineering Division of the Oak Ridge National Laboratory and has also taught in the Mathematics Department at the United States Naval Academy. His favorite part of teaching is connecting complex mathematical concepts to their applications. He finds most students are excited by math when they know how much of our modern world depends on it.

When he is not teaching (which is most of the time), Dr. Griffin's research interests are in applied mathematics, where he focuses on applied dynamical systems (especially on graphs), game theory, and optimization. His research has been funded by the National Science Foundation, the Office of Naval Research, the Army Research Office, the Intelligence Advanced Research Projects Agency, and the Defense Advanced Research Projects Agency. He has published over 100 peer-reviewed research papers in various forms of applied mathematics. His book, *Applied Graph Theory: An Introduction with Graph Optimization and Algebraic Graph Theory*, was also published by World Scientific in 2023.

Contents

Preface   vii

Acknowledgements   xiii

About the Author   xv

Part 1: Classical Game Theory   1

**1. Games Against the House with
   an Introduction to Probability Theory**   3

   1.1   Probability   3
   1.2   Random Variables and Expected Values   8
   1.3   Some Specialized Results on Probability   12
   1.4   Conditional Probability   15
   1.5   Independence   18
   1.6   Blackjack: A Game of Conditional Probability . . .   19
   1.7   The Monty Hall Problem and Decision Trees   22
   1.8   Bayes' Theorem   25
   1.9   Chapter Notes   27
   1.10   Exercises   29

2. Elementary Utility Theory   33

   2.1   Decision-Making Under Certainty   33
   2.2   Preference and the von Neumann–Morgenstern
          Assumptions   35
   2.3   Expected Utility Theorem   40
   2.4   Chapter Notes   46
   2.5   Exercises   48

3. Game Trees and Extensive Form   **51**

3.1   Graphs and Trees 51
3.2   Game Trees with Complete Information
       and No Chance 56
3.3   Game Trees with Incomplete Information 61
3.4   Games of Chance 65
3.5   Payoff Functions and Equilibria 66
3.6   Chapter Notes 81
3.7   Exercises 82

**4. Games and Matrices: Normal and
Strategic Forms**   **87**

4.1   Normal and Strategic Forms 87
4.2   Strategic-Form Games 89
4.3   Strategy Vectors and Matrix Games 93
4.4   Chapter Notes 95
4.5   Exercises 96

**5. Saddle Points, Mixed Strategies,
and Nash Equilibria**   **97**

5.1   Equilibria in Zero-Sum Games: Saddle Points . . . 98
5.2   Zero-Sum Games without Saddle Points 103
5.3   Mixed Strategies 105
5.4   Dominated Strategies and Nash Equilibria 111
5.5   The Indifference Theorem 117
5.6   The Minimax Theorem 120
5.7   Existence of Nash Equilibria 125
5.8   Finding Nash Equilibria in Simple Games 127
5.9   Nash Equilibria in General-Sum Games 131
5.10   Chapter Notes 134
5.11   Exercises 135

Part 2: Optimization and Game Theory   **137**

**6. An Introduction to Optimization and the
Karush–Kuhn–Tucker Conditions**   **139**

6.1   Motivating Example 139
6.2   A General Maximization Formulation 141

6.3 Gradients, Constraints, and Optimization 143
6.4 Convex Sets and Combinations 145
6.5 Convex and Concave Functions 147
6.6 Karush–Kuhn–Tucker Conditions 149
6.7 Relating Back to Game Theory 154
6.8 Chapter Notes 156
6.9 Exercises 157

7. **Linear Programming and Zero-Sum Games**   **159**

7.1 Linear Programs 160
7.2 Intuition on the Solution of Linear Programs 162
7.3 A Linear Program for Zero-Sum Game Players . . . 168
7.4 Solving Linear Programs Using a Computer 171
7.5 Standard Form, Slack and Surplus Variables 173
7.6 Optimality Conditions for Zero-Sum Games
    and Duality 175
7.7 Chapter Notes 183
7.8 Exercises 184

8. **Quadratic Programs and General-Sum Games**   **187**

8.1 Introduction to Quadratic Programming 187
8.2 Solving Quadratic Programming Problems
    Using Computers 188
8.3 General-Sum Games and Quadratic Programming . 189
8.4 Chapter Notes 202
8.5 Exercises 204

Part 3: Cooperation in Game Theory   **205**

9. **Nash's Bargaining Problem and
   Cooperative Games**   **207**

9.1 Payoff Regions in Two-Player Games 208
9.2 Collaboration and Multi-Criteria Optimization . . . 213
9.3 Nash's Bargaining Axioms 217
9.4 Nash's Bargaining Theorem 219
9.5 Chapter Notes 227
9.6 Exercises 228

10. An Introduction to *N*-Player Cooperative Games   229

10.1   Motivating Cooperative Games 229
10.2   Coalition Games 231
10.3   Division of Payoff to the Coalition 233
10.4   The Core 234
10.5   Shapley Values 237
10.6   Chapter Notes 239
10.7   Exercises 240

Appendix A. Introduction to Matrix Arithmetic   243

A.1   Matrices, Row and Column Vectors 243
A.2   Matrix Multiplication 245
A.3   Special Matrices and Vectors 246
A.4   Exercises 247

Appendix B. Essential Concepts from Vector Calculus   249

B.1   Geometry for Vector Calculus 249
B.2   Gradients 252
B.3   Exercises 254

Appendix C. Introduction to Evolutionary Games Using the Replicator Equation   257

C.1   Differential Equations 257
C.2   Fixed Points, Stability, and Phase Portraits 261
C.3   The Replicator Equation 264
C.4   Appendix Notes 271
C.5   Exercises 272

References   273

Index   281

Part 1
Classical Game Theory

Chapter 1

Games Against the House with an Introduction to Probability Theory

Chapter Goals: The goal of this chapter is to introduce probability theory and optimal decision-making in the context of casino games. By "casino game," we mean any game that has an element of chance and features a player (you) against the house (casino). Roulette, black jack, craps, etc., all fall into this category, as do many television game shows. Probability is introduced in a formal setting, as are random variables, expectation, and Bayes' theorem. A discussion of the *Monty Hall problem* concludes the chapter.

1.1 Probability

Remark 1.1. Our study of game theory begins with a characterization of optimal decision-making for an individual in the absence of any other players. The *games* we often see on television fall into this category. TV game shows (that do not pit players against each other in knowledge tests) often require a single player (who is, in a sense, playing against *the house*) to make a decision that will affect only her life.

Remark 1.2. In this chapter, we frequently discuss the TV game show *Deal or No Deal*. In this game, players are shown unmarked boxes[1] containing various amounts of money. At any given time, a banker offers the player an amount to quit the game. Players choose boxes, thus eliminating them. The objective is to retain (by luck) the box with the largest amounts of money.

Example 1.3. Congratulations! You have made it to the very final stage of *Deal or No Deal*. Two suitcases with money remain in play: One contains $0.01, while the other contains $1,000,000. The banker has offered you a payoff of $499,999 to quit. Do you accept the banker's safe offer or do you risk it all to try for $1,000,000. Suppose the banker offers you $100,000; what about $500,000 or $10,000?

Remark 1.4. Example 1.3 may seem contrived, but it has real-world implications and most of the components needed for a serious discussion of decision-making under risk. In order to study these concepts formally, we need a grounding in probability theory. Unfortunately, a formal study of probability requires a heavy dose of measure theory, which is well beyond the scope of an introductory course on game theory. Therefore, the following definitions are meant to be intuitive rather than mathematically rigorous.

Definition 1.5 (Outcome). Let Ω be a finite set of elements describing the outcome of a chance event (a coin toss, a roll of the dice, etc.). We call Ω the *sample space*. Each element of Ω is called an *outcome*.

Example 1.6. In the case of Example 1.3, the only thing we care about is the position of $1,000,000 and $0.01 within the boxes. In this case, Ω consists of two possible outcomes: either $1,000,000 is in box number 1 (while $0.01 is in box number 2) or $1,000,000 is in box number 2 (while $0.01 is in box number 1).

Formally, let us refer to the first outcome as A and the second outcome as B. Then, $\Omega = \{A, B\}$.

Definition 1.7 (Event). If Ω is a sample space, then an event is any subset of Ω. We write this as $E \subseteq \Omega$ to indicate that event E is a subset of Ω. If we are certain that E is a proper subset (i.e., $E \neq \Omega$), we write $E \subset \Omega$.

---

[1]Some versions of the show use suitcases, others use boxes.

Example 1.8. The sample space in Example 1.3 consists of precisely four events: $\emptyset$ (the empty event), $\{A\}$, $\{B\}$, and $\{A, B\} = \Omega$. These four sets represent all possible subsets of the set $\Omega = \{A, B\}$.

Definition 1.9 (Union). If $E, F \subseteq \Omega$ are both events, then $E \cup F$ is the *union* of the sets E and F and consists of all outcomes in either E or F. Event $E \cup F$ occurs if event E or event F occurs.

Example 1.10. Consider the roll of a fair six-sided die. The outcomes are $\Omega = \{1, \ldots, 6\}$. If $E = \{1, 3\}$ and $F = \{2, 4\}$, then $E \cup F = \{1, 2, 3, 4\}$ and will occur as long as we don't roll a 5 or 6.

Definition 1.11 (Intersection). If $E, F \subseteq \Omega$ are both events, then $E \cap F$ is the *intersection* of the sets E and F and consists of all outcomes in both E and F. Event $E \cap F$ occurs if both event E and event F occur.

Example 1.12. Again, consider the roll of a fair six-sided die. The outcomes are $1, \ldots, 6$. If $E = \{1, 2\}$ and $F = \{2, 4\}$, then $E \cap F = \{2\}$ and will occur only if we roll a 2.

Definition 1.13 (Mutual Exclusivity). Two events $E, F \subseteq \Omega$ are said to be *mutually exclusive* if and only if $E \cap F = \emptyset$.

Definition 1.14 (Discrete Probability Distribution Function). Given a discrete sample space Ω, let $\mathcal{F}$ be the set of all events on Ω. A *discrete probability function* is a mapping $P : \mathcal{F} \to [0, 1]$ with the following properties:

(1) $P(\Omega) = 1$; and
(2) if $E, F \in \mathcal{F}$ and $E \cap F = \emptyset$, then $P(E \cup F) = P(E) + P(F)$.

Remark 1.15 (Power Set). In this definition, we consider the set $\mathcal{F}$ as the set of all events over a set of outcomes Ω. This is an example of the *power set*: the set of all subsets of a set. We sometimes denote this set as 2^Ω. Thus, if Ω is a set, then 2^Ω is the power set of Ω or the set of all subsets of Ω.

Remark 1.16. Definition 1.14 is surprisingly technical and probably does not conform to your ordinary sense of what probability is. It's best not to think of probability in this very formal way. Instead, it suffices to think that a probability function assigns a number to an outcome (or event) that tells you the chances of it occurring.

Put more simply, suppose we could run an experiment where the result of that experiment will be an outcome in Ω. Then, the function P simply tells us the proportion of times we will observe an event $E \subset \Omega$ if we run this experiment an exceedingly large number of times.

Example 1.17. Suppose we could play the *Deal or No Deal* example over and over again and observe where the money ends up. A smart game show would mix the money up so that approximately one-half of the time we observe \$1,000,000 in Suitcase 1, and the other half of the time we observe \$1,000,000 in Suitcase 2.

A probability distribution formalizes this notion and might assign $1/2$ to event $\{A\}$ and $1/2$ to event $\{B\}$. However, to obtain a true probability distribution, we must also assign probabilities to $\emptyset$ and $\{A, B\}$. In the former case, we know that something must happen. Therefore, we can assign 0 to the event $\emptyset$. In the latter case, we know for certain that either outcome A or B must occur, and so, in this case, we assign a value of 1 to $P(\Omega)$.

Example 1.18. In a fair six-sided die, the probability of rolling any value is $1/6$. Formally, $\Omega = \{1, 2, \ldots, 6\}$, and any roll is an event with only one element: $\{\omega\}$, where ω is some value in Ω. If we consider the event $E = \{1, 2, 3\}$, then $P(E)$ gives us the probability that we will roll a 1, 2, or 3. Since $\{1\}$, $\{2\}$, and $\{3\}$ are disjoint sets and $\{1, 2, 3\} = \{1\} \cup \{2\} \cup \{3\}$, we know that

$$P(E) = \frac{1}{6} + \frac{1}{6} + \frac{1}{6} = \frac{1}{2}.$$

Definition 1.19 (Discrete Probability Space). The triple $(\Omega, \mathcal{F}, P)$ is called a *discrete probability space* over Ω.

Definition 1.20 (Set of All Probability Spaces). Let Ω be a discrete sample space. The set of all probability spaces defined on Ω is denoted $\Delta(\Omega)$.

Remark 1.21. The previous definition is used in our study of general utility formulations. In the case when $\Omega = \{\omega_1, \ldots, \omega_n\}$, we can also think of Δ as being in one-to-one correspondence with the set of all points $(p_1, \ldots, p_n) \in \mathbb{R}^n$ such that:

(1) $p_1 + p_2 + \cdots + p_n = 1$ and
(2) $p_i \geq 0$.

That is, if $(\Omega, \mathcal{F}, P) \in \Delta(\Omega)$, then $P(\omega_i) = p_i$. We will encounter the set

$$\Delta_n = \left\{ (p_1, \ldots, p_n) \in \mathbb{R}^n : \sum_i p_i = 1, \ p_i \geq 0, \ i \in \{1, \ldots, n\} \right\}$$

again in Chapter 5.

Lemma 1.22. *Let $(\Omega, \mathcal{F}, P)$ be a discrete probability space. Then, $P(\emptyset) = 0$.*

Proof.   The sets $\Omega \in \mathcal{F}$ and $\emptyset \in \mathcal{F}$ are disjoint (i.e., $\Omega \cap \emptyset = \emptyset$). Thus,

$$P(\Omega \cup \emptyset) = P(\Omega) + P(\emptyset).$$

We know that $\Omega \cup \emptyset = \Omega$. Thus, we have

$$P(\Omega) = P(\Omega) + P(\emptyset) \implies 1 = 1 + P(\emptyset) \implies 0 = P(\emptyset).$$

$\square$

Lemma 1.23. *Let $(\Omega, \mathcal{F}, P)$ be a discrete probability space, and let $E, F \in \mathcal{F}$. Then,*

$$P(E \cup F) = P(E) + P(F) - P(E \cap F). \tag{1.1}$$

Proof.   If $E \cap F = \emptyset$, then by definition, $P(E \cup F) = P(E) + P(F)$ but $P(\emptyset) = 0$; therefore, $P(E \cup F) = P(E) + P(F) - P(E \cap F)$.

Suppose $E \cap F \neq \emptyset$. Then, let

$$E' = \{\omega \in E | \omega \notin F\} \quad \text{and}$$

$$F' = \{\omega \in F | \omega \notin E\}.$$

This is illustrated in Fig. 1.1. Then, we know the following:

(1) $E' \cap F' = \emptyset$,
(2) $E' \cap (E \cap F) = \emptyset$,
(3) $F' \cap (E \cap F) = \emptyset$,
(4) $E = E' \cup (E \cap F)$, and
(5) $F = F' \cup (E \cap F)$.

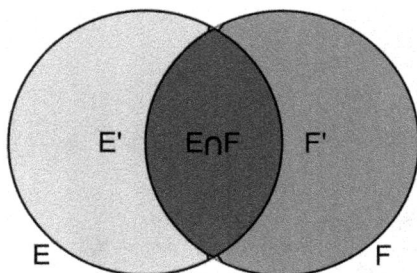

Fig. 1.1.   An illustration of the probabilities used in the proof of Lemma 1.23.

Thus (by the inductive extension of Property 2 in Definition 1.14), we know that

$$P(E \cup F) = P(E' \cup F' \cup (E \cap F)) = P(E') + P(F') + P(E \cap F).$$

$$(1.2)$$

We also know that

$$P(E) = P(E') + P(E \cap F) \implies P(E') = P(E) - P(E \cap F) \quad (1.3)$$

and

$$P(F) = P(F') + P(E \cap F) \implies P(F') = P(F) - P(E \cap F). \quad (1.4)$$

Combining these three equations yields

$$P(E \cup F) = P(E) - P(E \cap F) + P(F) - P(E \cap F) + P(E \cap F)$$
$$= P(E \cup F) = P(E) + P(F) - P(E \cap F). \quad (1.5)$$

This completes the proof.   □

1.2   Random Variables and Expected Values

Remark 1.24. Intuitively, a random variable is a variable X whose value is not known *a priori* and which is determined according to some probability distribution P that is part of a probability space $(\Omega, \mathcal{F}, P)$. Naturally, we can make this more rigorous, but it is not necessary.

Example 1.25. Suppose that we consider flipping a fair coin. Then, the probability of seeing *heads* (or *tails*) should be $1/2$. If we let X be a random variable that provides the outcome of the flip, then it will take on a value of either *heads* or *tails*, and it will take each value half of the time (in the long run).

Remark 1.26. The problem with allowing a random variable to take on arbitrary values (like *heads* or *tails*) is that it makes it difficult to use random variables in formulas involving numbers. There is a *very* technical definition of a random variable that arises in formal probability theory. However, it is well beyond the scope of this book. We can, however, get a flavor of this definition in the following restricted form.

Definition 1.27. Let $(\Omega, \mathcal{F}, P)$ be a discrete probability space. Let $D \subseteq \mathbb{R}$ be a finite discrete subset of the real numbers. A random variable X is a function that maps each element of Ω to an element of D. Formally, $X : \Omega \to D$.

Remark 1.28. Clearly, if $S \subseteq D$, then $X^{-1}(S) = \{\omega \in \Omega | X(\omega) \in S\} \in \mathcal{F}$. We can think of the probability of X taking on a value in $S \subseteq D$ as precisely $P(X^{-1}(S))$.

Using this observation, if $(\Omega, \mathcal{F}, P)$ is a discrete probability distribution function, $X : \Omega \to D$ is a random variable, and $x \in D$, then let $P(x) = P(X^{-1}(\{x\}))$. That is, the probability of X taking the value x is the probability of the element in Ω corresponding to x.

Example 1.29. Consider our coin-flipping random variable. Instead of having X take a value of either *heads* or *tails*, we can instead let X take on value of 1 if the coin comes up *heads* and 0 if the coin comes up *tails*. Thus, if $\Omega = \{heads, tails\}$, then $X(heads) = 1$ and $X(tails) = 0$.

Example 1.30. When Ω is already a subset of $\mathbb{R}$, then defining random variables is easy. The random variable can simply be the obvious mapping from Ω into itself. For example, if we consider rolling a fair die, then $\Omega = \{1, \ldots, 6\}$, and any random variable defined on $(\Omega, \mathcal{F}, P)$ will take on values $1, \ldots, 6$.

Definition 1.31 (Expected Value). Let $(\Omega, \mathcal{F}, P)$ be a discrete probability distribution, and let $X : \Omega \to D$ be a random variable.

Then, the *expected value* of X is

$$\mathbb{E}(X) = \sum_{x \in D} xP(x). \tag{1.6}$$

Example 1.32. Let's play a die-rolling game. You put up your own money. Even numbers lose \$10 times the number rolled, while odd numbers win \$12 times the number rolled. What is the expected amount of money you will win in this game?

Let $\Omega = \{1, \ldots, 6\}$. Then, $D = \{12, -20, 36, -40, 60, -60\}$; these are the dollar values you will win for various dice outcomes. Then, the expected value of X is

$$\mathbb{E}(X) = 12 \left(\frac{1}{6}\right) + (-20) \left(\frac{1}{6}\right)$$

$$+ 36 \left(\frac{1}{6}\right) + (-40) \left(\frac{1}{6}\right) + 60 \left(\frac{1}{6}\right) + (-60) \left(\frac{1}{6}\right) = -2.$$

Would you still want to play this game considering the expected payoff is $-\$2$?

Example 1.33 (Roulette). A roulette wheel consists of 38 pockets (slots) numbered 0–36, with an extra pocket labeled 00. Pockets 0 and 00 are green. The remaining pockets are black or red. Eighteen are black, and eighteen are red. You bet by placing chips on a board that has the numbers arranged in rows and columns. Depending on the bet you are making the sample space may change. For example, if you are betting on color, the sample space is $\Omega_{\text{color}} = \{Red, Black, Green\}$. If you are betting strictly on numbers, the sample space is $\Omega_{\text{number}} = \{00, 0, 1, \ldots, 36\}$. A representation of a roulette board and wheel is shown in Fig. 1.2. Payoffs in roulette are given in ratios. For example, a bet on any number gives a payout of 35 to 1. That means if you bet \$1 on number 1 and win, you get the original \$1 back and \$35 additional dollars. The payoff ratio establishes the mapping $X : \Omega \to \mathbb{R}$, which is the *random variable* (payoff) in this case.

Ignoring the notational complexity, on a single-number bet, the expected profit is

$$35 \cdot \left(\frac{1}{38}\right) - 1 \cdot \left(\frac{37}{38}\right) = -\frac{1}{19} \approx -0.053. \tag{1.7}$$

More complex betting strategies are possible, but they lead to similar results. Playing red returns a payout of 1-to-1, giving an expected

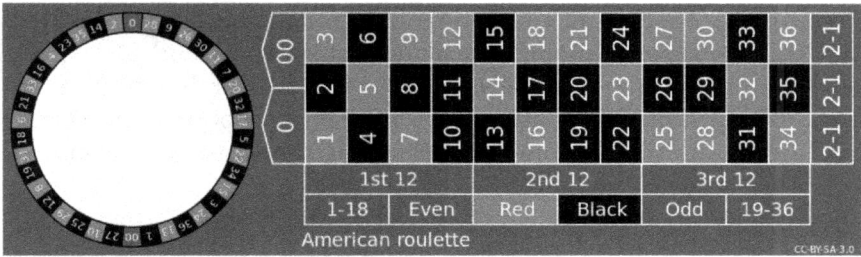

Fig. 1.2. An (American) roulette wheel is shown above. A French roulette wheel lacks the 00 pocket. This image was obtained from https://commons.wikimedia. org/wiki/File:American_roulette.svg under CC-BY-SA-3.0 license.

profit of

$$1 \cdot \left(\frac{18}{38}\right) - 1 \cdot \left(\frac{20}{38}\right) = -\frac{1}{19} \approx -0.053. \tag{1.8}$$

This is because there are 18 pockets that are red and 20 pockets that are not red (18 black and 2 green).

Example 1.34 (Nick the Greek). Nick the Greek was a *professional* gambler in Las Vegas during the 1940s. Though he died poor (due to poker losses) while in Las Vegas, he developed a system in which he would gamble against the players at the table and not the games themselves. For example, if a person was convinced that the roulette wheel was going to come up red, he (Nick) might say, "I'll give you 1-to-1 odds that it's not going to be red. If you're right, you'll double your bet!" People's superstitions would work against them. If the person lost the spin, Nick got, say, $1. If they won the spin, e.g., Nick lost $1. In effect, Nick became the house. Now, on a red bet, Nick's expected payoff would be

$$-1 \cdot \left(\frac{18}{38}\right) + 1 \cdot \left(\frac{20}{38}\right) = \frac{1}{19}. \tag{1.9}$$

That means, in the long run, Nick would come out ahead on all his side bets. Since Nick was wealthy to begin with (his wealth was partially inherited), he could cover small losses periodically and reap the marginal long-run gain. In essence, Nick was acting like a one-man hedge fund.

Example 1.35. Return to Example 1.3. You are on *Deal or No Deal*, and the two boxes remaining in play contain \$0.01 and \$1,000,000, but you do not know which is which. The banker has offered you \$499,999 to quit. You have a 50% chance of choosing the correct box. Let W be the random variable giving the winnings if you play. It has an expected value of

$$\mathbb{E}(W) = \left(\frac{1}{2}\right) 1,000,000 + \left(\frac{1}{2}\right) 0.01 = 500,000.005.$$

This value is larger than the banker's offer. So, one school of thought argues you should play on.

However, here is another way to look at the problem. The banker's money is yours with no risk; suppose you consider it already your money. Rather than assuming you started with nothing, assume you're starting with \$499,999. Let G be the random variable denoting your gain from deciding to play. Then, we have

$$\mathbb{E}(G) = \frac{1}{2}(0.01 - 499,999) + \frac{1}{2}(500,000 - 499,999) = -249998.995.$$

Viewed in this way, the banker's offer seems to be a good deal. The decision to take the banker's offer has no risk, and you keep the \$499,999. The decision to play on is risky and, viewed from the perspective of gain, will likely cost you almost a quarter of a million dollars.

This example illustrates a fundamental problem in game theory and decision theory: Determining the objective or payoff function in complex situations is challenging. In Chapter 2, we discuss von Neumann and Morgenstern's [5] approach to this problem.

1.3   Some Specialized Results on Probability

Remark 1.36. This section is optional. The results herein are used only once in Chapter 2.

Remark 1.37. The proof of the following lemma is left as an exercise.

Lemma 1.38. *Let $(\Omega, \mathcal{F}, P)$ be a discrete probability space, and let $E, F \in \mathcal{F}$. Then,*

$$P(E) = P(E \cap F) + P(E \cap F^c). \tag{1.10}$$

Remark 1.39. There are several ways to prove the following lemma. We choose a classic set-theoretic proof.

Lemma 1.40. *Let Ω be a set (sample space), and suppose that $E, F_1, \ldots, F_n$ are subsets of Ω. Then,*

$$E \cap \bigcup_{i=1}^{n} F_i = \bigcup_{i=1}^{n} (E \cap F_i). \tag{1.11}$$

That is, intersection distributes over union.

Proof.   We prove set containment in both directions. That is, we prove that

$$E \cap \bigcup_{i=1}^{n} F_i \subseteq \bigcup_{i=1}^{n} (E \cap F_i) \quad \text{and}$$

$$E \cap \bigcup_{i=1}^{n} F_i \supseteq \bigcup_{i=1}^{n} (E \cap F_i).$$

For simplicity, define

$$F = \bigcup_{i=1}^{n} F_i.$$

Suppose we have $\omega \in E \cap F$. Then, $\omega \in E$ and $\omega \in F$. That means there is at least one i so that $\omega \in F_i$. Then, $\omega \in E \cap F_i$; consequently,

$$\omega \in \bigcup_{i=1}^{n} (E \cap F_i).$$

Now, we proceed in the opposite direction. Suppose that

$$\omega \in \bigcup_{i=1}^{n} (E \cap F_i).$$

Then, there is at least one i so that $\omega \in E$ and $\omega \in F_i$. This implies that $\omega \in E \cap F$. This completes the proof. $\square$

Theorem 1.41. *Let $(\Omega, \mathcal{F}, P)$ be a discrete probability space, and let $E \in \mathcal{F}$. Let $F_1, \ldots, F_n$ be any pairwise disjoint collection of sets that partition Ω. That is, assume*

$$\Omega = \bigcup_{i=1}^{n} F_i, \tag{1.12}$$

and $F_i \cap F_j = \emptyset$ if $i \neq j$. Then,

$$P(E) = \sum_{i=1}^{n} P(E \cap F_i). \tag{1.13}$$

Proof.   We proceed by induction on n. If $n = 1$, then $F_1 = \Omega$, and we know that $P(E) = P(E \cap \Omega)$ by necessity. Therefore, suppose the statement is true for $k \leq n$. We show that the statement is true for $n + 1$.

Let $F_1, \ldots, F_{n+1}$ be pairwise disjoint subsets satisfying Eq. (1.12). Let

$$F = \bigcup_{i=1}^{n} F_i. \tag{1.14}$$

Clearly, if $x \in F$, then $x \notin F_{n+1}$ since $F_{n+1} \cap F_i = \emptyset$ for $i = 1, \ldots, n$. Also, if $x \notin F$, then $x \in F_{n+1}$ since from Eq. (1.12), we must have $F \cup F_{n+1} = \Omega$. Thus, $F^c = F_{n+1}$, and we can conclude inductively that

$$P(E) = P(E \cap F) + P(E \cap F_{n+1}). \tag{1.15}$$

We may apply Lemma 1.40 to show that

$$E \cap F = E \cap \bigcup_{i=1}^{n} F_i = \bigcup_{i=1}^{n} (E \cap F_i). \tag{1.16}$$

Note that if $i \neq j$, then $(E \cap F_i) \cap (E \cap F_j) = \emptyset$ because $F_i \cap F_j = \emptyset$; therefore,

$$P(E \cap F) = P\left(\bigcup_{i=1}^{n}(E \cap F_i)\right) = \sum_{i=1}^{n} P(E \cap F_i). \tag{1.17}$$

Thus, we may write

$$P(E) = \sum_{i=1}^{n} P(E \cap F_i) + P(E \cap F_{n+1}) = \sum_{i=1}^{n+1} P(E \cap F_i). \tag{1.18}$$

This completes the proof.   □

Example 1.42. In the casino game craps, we roll two dice, and winning combinations are determined by the sum of the values on the dice. An ideal first craps roll is 7. The sample space Ω in which we are interested has 36 elements, one each for the possible values the dice will show (the related set of sums can be easily obtained).

Suppose that the dice are colored blue and red (so they can be distinguished). Let's suppose we are interested in the event that we roll 1 on the blue die and that the pair of values obtained sums to 7. There is only one way this can occur, namely, we roll 1 on the blue die and 6 on the red die. Thus, the probability of this occurring is $\frac{1}{36}$. In this case, event E is the event that we roll 7 in our craps game, and event F_1 is the event that the blue die shows a 1. We could also consider the event F_2, in which the blue die shows a 2. By similar reasoning, we know that the probability of both E and F_2 occurring is $\frac{1}{36}$. In fact, if F_i is the event that the blue die shows a value of i $(i = 1, \ldots, 6)$, then we know that

$$P(E \cap F_i) = \frac{1}{36}.$$

Clearly, the events F_i $(i \in \{1, \ldots, 6\})$ are pairwise disjoint (you cannot have both 1 and 2 on the same die). Furthermore, $\Omega = F_1 \cup F_2 \cup \cdots \cup F_6$ because some number has to come up on the blue die. Thus, we can compute

$$P(E) = \sum_{i=1}^{6} P(E \cap F_i) = \frac{6}{36} = \frac{1}{6},$$

which is precisely what we would expect. The probability of rolling 7 with two dice is $\frac{1}{6}$.

1.4   Conditional Probability

Remark 1.43. Suppose we are given a discrete probability space, $(\Omega, \mathcal{F}, P)$, and we are told that an event E has occurred. We now wish to compute the probability that some other event F will (or has) occurred. This value is called the conditional probability of event F given event E and is written $P(F|E)$.

Example 1.44. Consider an experiment where we roll a fair six-sided die twice. The sample space in this case is the set $\Omega = \{(x,y)|x = 1,\ldots,6,\ y = 1,\ldots,6\}$. Suppose I roll a 2 on the first try. I want to know what the probability of rolling a combined score of 8 is. That is, given that I have rolled a 2, I wish to determine the conditional probability of rolling a 6.

Since the die is fair, the probability of rolling any pair of values $(x,y) \in \Omega$ is equally likely. There are 36 elements in Ω, and so each is assigned a probability of $1/36$. That is, $(\Omega, \mathcal{F}, P)$ is defined so that $P((x,y)) = 1/36$ for each $(x,y) \in \Omega$.

Let E be the event that we roll a 2 on the first try. We wish to assign a new set of probabilities to the elements of Ω to reflect the fact that the event E has occurred. We know that our final outcome must have the form $(2,y)$, where $y \in \{1,\ldots,6\}$. **In essence, E becomes our new sample space.** Further, we know that each outcome with form $(2,y)$ is equally likely because the dice are fair. Thus, we may assign

$$P[(2,y)|E] = \frac{1}{6}$$

for each $y \in \{1,\ldots,6\}$ and $P[(x,y)|E] = 0$ just in case $x \neq 2$. That is, any outcome (x,y) that is not in E is assigned a probability of 0. We do this because we know that we have already observed the number 2 on the first roll, so it's impossible to see a first number not equal to 2.

At last, we can answer the question we originally posed. The only way to obtain a sum equal to 8 is to roll a 6 on the second attempt. Thus, the probability of rolling a combined score of 8, given that we roll 2 on the first roll, is $\frac{1}{6}$.

Lemma 1.45. *Let $(\Omega, \mathcal{F}, P)$ be a discrete probability space, and suppose that event $E \subseteq \Omega$. Then, $(E, \mathcal{F}_E, P_E)$ is a discrete probability space, with*

$$P_E(F) = \frac{P(F)}{P(E)} \tag{1.19}$$

for all $F \subseteq E$ and $P_E(\omega) = 0$ for any $\omega \notin E$.

Proof. Our objective is to show that $(E, \mathcal{F}_E, P_E)$ is a properly defined probability space.

If $\omega \notin E$, then we can assign $P_E(\omega) = 0$. Suppose that $\omega \in E$. For $(E, \mathcal{F}_E, P_E)$ to be a discrete probability space, we must have $P_E(E) = 1$, or

$$P_E(E) = \sum_{\omega \in E} P_E(\omega) = 1. \tag{1.20}$$

We know from Definition 1.14 that

$$P(E) = \sum_{\omega \in E} P(\omega).$$

Thus, if we assign $P_E(\omega) = P(\omega)/P(E)$ for all $\omega \in E$, then Eq. (1.20) will be satisfied automatically. Since for any $F \subseteq E$ we know that

$$P(F) = \sum_{\omega \in F} P(\omega),$$

it follows that $P_E(F) = P(F)/P(E)$. Finally, if $F_1, F_2 \subseteq E$ and $F_1 \cap F_2 = \emptyset$, then the fact that $P_E(F_1 \cup F_2) = P_E(F_1) + P_E(F_2)$ follows from the properties of the original probability space $(\Omega, \mathcal{F}, P)$. Thus, $(E, \mathcal{F}_E, P_E)$ is a discrete probability space. $\qquad \square$

Remark 1.46. The previous lemma gives us a direct way to construct $P(F|E)$ for arbitrary $F \subseteq \Omega$. Clearly, if $F \subseteq E$, then

$$P(F|E) = P_E(F) = \frac{P(F)}{P(E)}.$$

Now, suppose that F is not a subset of E but that $F \cap E \neq \emptyset$. Then, clearly, the only possible events that can occur in F, given that E has occurred, are the ones that are also in E. Thus, $P_E(F) = P_E(E \cap F)$. More to the point, we have

$$P(F|E) = P_E(F \cap E) = \frac{P(F \cap E)}{P(E)}. \tag{1.21}$$

This leads to the following definition.

Definition 1.47 (Conditional Probability). Given a discrete probability space $(\Omega, \mathcal{F}, P)$ and an event $E \in \mathcal{F}$, the conditional probability of event $F \in \mathcal{F}$ given event E is

$$P(F|E) = \frac{P(F \cap E)}{P(E)}. \tag{1.22}$$

1.5   Independence

Definition 1.48 (Independence). Let $(\Omega, \mathcal{F}, P)$ be a discrete probability space. Two events $E, F \in \mathcal{F}$ are called *independent* if $P(E|F) = P(E)$ and $P(F|E) = P(F)$.

Theorem 1.49. *Let $(\Omega, \mathcal{F}, P)$ be a discrete probability space. If $E, F \in \mathcal{F}$ are independent events, then $P(E \cap F) = P(E)P(F)$.*

Proof.   We know that

$$P(E|F) = \frac{P(E \cap F)}{P(F)} = P(E).$$

Multiplying by $P(F)$, we obtain $P(E \cap F) = P(E)P(F)$. This completes the proof.   □

Example 1.50. Consider rolling a fair die twice in a row. Let Ω be the sample space of possible results. Thus,

$$\Omega = \{(x, y) | x = 1, \ldots, 6, \ y = 1, \ldots, 6\}.$$

Let E be the event that we obtain a 6 on the first roll. Then,

$$E = \{(6, y) : y = 1, \ldots, 6\},$$

and let F be the event that we obtain a 6 on the second roll, so that

$$F = \{(x, 6) : x = 1, \ldots, 6\}.$$

These two events are independent. The first roll *cannot* affect the outcome of the second roll, thus $P(F|E) = P(F)$. We know that $P(E) = P(F) = \frac{1}{6}$. That is, there is a one in six chance of observing a 6. Thus, the chance of rolling double sixes in two rolls is precisely the probability of both events E and F occurring. Using our result on independent events, we can see that

$$P(E \cap F) = P(E)P(F) = \left(\frac{1}{6}\right)^2 = \frac{1}{36},$$

just as we expect it to be.

Example 1.51. Suppose we are interested in the probability of rolling at least one 6 in two rolls of a single die. Again, the rolls are independent. Let's consider the probability of not rolling a 6 at all. Let E be the event that we *do not* roll a 6 in the first roll. Then, $P(E) = 5/6$ (as there are five ways to not roll a 6). If F is the event that we do not roll a 6 on the second roll, then again $P(F) = 5/6$. Since these events are independent (as before), we can compute $P(E \cap F) = (5/6)(5/6) = 25/36$. This is the probability of not rolling a 6 on the first roll and not rolling a 6 on the second roll. We are interested in rolling at least one 6. Thus, if G is the event of not rolling a 6 at all, then G^c must be the event of rolling at least one 6. Thus, $P(G^c) = 1 - P(G) = 1 - 25/36 = 11/36$.

1.6 Blackjack: A Game of Conditional Probability

Remark 1.52. In the card game blackjack, a dealer deals cards to players and himself or herself. Non-face cards have the value shown; i.e., a two-card is worth the value of two. Face cards all have a value of 10, except aces, which are either worth 11 or 1, depending on what is best for the player. The object is to continue requesting cards (by saying "hit me") to achieve a total value as close to 21 as possible without going over. A player who wishes to receive no more cards is said to be "standing" or "standing pat." If a player achieves a value closer to 21 than the dealer, then he/she wins and doubles his/her bet. Otherwise, the dealer (house) wins. There are several nuances to the game (doubling down, insurance, etc.) that are outside the scope of this discussion. See Ref. [6] for additional details.

When blackjack is played with a dealer and a single player, it is a simple game against the house since the dealer is required to follow specific rules with respect to his/her cards. As a result, there is only one decision-maker (the solo player), as opposed to a game like poker in which there are multiple decision-makers (players). It is interesting to note that multi-player blackjack admits only a weak coupling between the players and the rewards they receive because all players share the same deck(s) used by the dealer. Even in this case, all players play directly against the house and not against each other. Player decisions only affect the *conditional probability distribution* on the next card. As such, blackjack (and similar games) admit certain strategies that can help players improve their chances of winning. These are usually called card-counting strategies.

Table 1.1.   A table of three example card counting strategies. The level of the count is the number of different distinct non-zero values that can be assigned to a card.

Strategy	2	3	4	5	6	7	8	9	10/Face	A	Level
Hi-Lo	+1	+1	+1	+1	+1	0	0	0	−1	−1	1
Hi-Opt I	0	+1	+1	+1	+1	0	0	0	−1	0	1
Hi-Opt II	+1	+1	+2	+2	+1	+1	0	0	−2	0	2

Card counting in blackjack is designed to allow the player to determine when the house has an advantage (without computing conditional probabilities) so that he/she can adjust the betting strategy (amount) accordingly. As cards are shown, the player adjusts the count according to a table of values for different cards. Three examples of card-counting strategies are shown in Table 1.1. Additional strategies can be found in Ref. [7]. In a counting strategy, the level of the count is the number of different distinct non-zero values that can be assigned to a card.

High counts favor the player, while low counts favor the dealer. This is (in some sense) intuitively clear. The more high cards that are seen, the lower the count goes and the less likely a player will see a large initial value on a deal. When many low cards are in play, players face the risk of busting (going over 21) as they try to increase the value of the hand dealt.

Example 1.53. We use a simple blackjack example to illustrate card counting. Suppose you are sitting at the blackjack table, and you see the configuration shown in Fig. 1.3. Assume you have already observed the following cards in a previous round of play: $A\heartsuit$, $J\spadesuit$, $6\clubsuit$, $2\heartsuit$, $5\diamondsuit$, $10\clubsuit$, $8\diamondsuit$, and $6\spadesuit$. You must decide whether to hit or not. Assuming you are playing with a 52-card deck, deciding whether to hit or not can be accomplished by computing the conditional probability of going over 21 (busting) given the current state of the cards and whether you hit or not. In this case, you will not bust if you get a 4 or less. Given the cards that have already been drawn, we see that there are 37 cards remaining in the deck. Of the 15 cards that have been drawn, you know the value of 13 of them. This implies the following:

Dealer

$$\boxed{K\spadesuit} \quad \boxed{??}$$

$$\boxed{Q\clubsuit} \quad \boxed{3\heartsuit} \quad \boxed{??} \qquad \boxed{7\diamondsuit} \quad \boxed{K\heartsuit}$$

Other Player   **You**

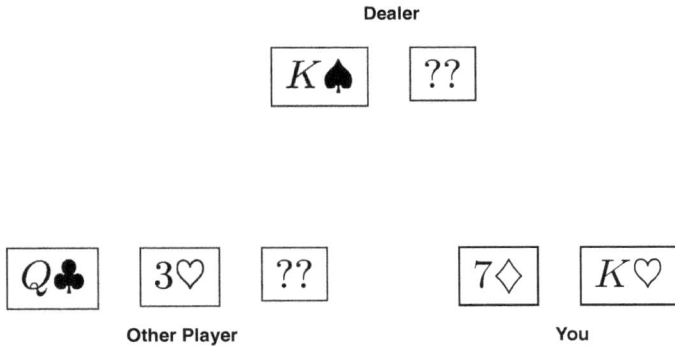

Fig. 1.3. You are sitting at a blackjack table. The dealer holds a king and something. You hold a 7 and a king. Do you hit?

- There are at most four 4's remaining in the deck.
- There are at most three 3's remaining in the deck.
- There are at most three 2's remaining in the deck.
- There are at most three A's remaining in the deck.

Therefore, there are at most 13 cards out of the remaining 37 that will help you. A rough back-of-the-envelope computation tells you that

$$\Pr(\text{Drawing a card with value} \le 4) \le \frac{13}{37} \approx 0.35. \tag{1.23}$$

Thus, the probability of busting is at least 65%. Consequently, you might consider holding with your 17, even though the dealer is showing a king. The probability we have just estimated is the conditional probability of drawing a card with a value of 4 or less, given the current state of the cards in play. The event we conditioned on is the current set of cards that have been drawn, or, equivalently, the cards remaining in the deck.

Let's investigate what a simple hi-lo count says about this situation. Using the counting system, we see that we have a count of

$$-1(A\heartsuit) - 1(J\spadesuit) + 1(6\clubsuit) + 1(2\heartsuit) + 1(5\diamondsuit) - 1(10\clubsuit) + 0(8\diamondsuit)$$
$$+1(6\spadesuit) - 1(K\spadesuit) - 1(Q\clubsuit) + 1(3\heartsuit) + 0(7\diamondsuit) - 1(K\heartsuit) = -1.$$

This suggests a slight advantage for the dealer at this point. In the next round of play, the count suggests that some caution be exercised in placing a bet, which is consistent with our observations thus far. At the moment, the probability of busting is quite high.

It is worth noting that the count is about the next set of bets. It's not strictly about what to do next in terms of hitting, though it certainly can inform that decision. In this case, the odds favor standing pat in order to avoid going over 21, which is a guaranteed loss.

1.7   The Monty Hall Problem and Decision Trees

Remark 1.54. The game show *Let's Make a Deal* originated in the United States in 1963 [8] and was originally hosted by Monty Hall. The show features the host offering various deals and trades to contestants, who often have incomplete information.

Example 1.55 (The Monty Hall Problem). You are a contestant on *Let's Make a Deal.* You must choose between Door Number 1, Door Number 2, and Door Number 3. Behind one of these doors is a fabulous prize. Behind the other two doors are goats. Once you choose your door, the host, Monty Hall, will reveal a door that does not have a big deal. At this point, you can decide if you want to keep the original door you chose or switch doors. When the time comes, what do you do?

It is tempting at first to suppose that it doesn't matter whether you switch or not. You have a 1/3 chance of choosing the correct door on your first try, so why would that change after you are provided information about an incorrect door? It turns out that it does matter.

To solve this problem, it helps to understand the set of potential outcomes and the information associated with the decision. There are really three possible pieces of information that determine an outcome:

(1) which door the producer chooses for the big deal,
(2) which door you choose first, and
(3) whether you switch or not.

For the first decision, there are three possibilities (three doors). For the second decision, there are again three possibilities (again, three doors). For the third decision, there are two possibilities (you either switch or not). Thus, there are $3 \times 3 \times 2 = 18$ possible outcomes. These outcomes can be visualized in the order in which the decisions are made (more or less), which is shown in Fig. 1.4. The first step (where the producers choose a door to hide the prize) is not observable by the contestant, so we adorn this part of the diagram with a box. When we discuss *game trees* in Chapter 3, we explain this notation more completely.

The next to last row (labeled "Switch") of Fig. 1.4 illustrates the 18 elements of the probability space. We assume that they are all equally likely (i.e., you randomly choose a door, you randomly decide to switch, and the producers of the show randomly choose a door for hiding the prize). In this case, the probability of any outcome is 1/18. Now, let's focus exclusively on the outcomes in which we decide to switch. In Fig. 1.4, these appear with bold, black borders. This is the conditioning event, that is, the event set E. Let the event F consist of those outcomes for which the contestant wins. This is shown in the bottom row of Fig. 1.4, with a W. We are interested in $P(F|E)$. That is, what are the chances of winning, given that we actively choose to switch?

Within E, there are precisely 6 outcomes in which we win. If each of these mutually exclusive outcomes has a probability of 1/18,

$$P(E \cap F) = 6\left(\frac{1}{18}\right) = \frac{1}{3}.$$

Obviously, we switch in 9 of the 18 possible independent outcomes, so

$$P(E) = 9\left(\frac{1}{18}\right) = \frac{1}{2}.$$

Thus, we can compute

$$P(F|E) = \frac{P(E \cap F)}{P(E)} = \frac{1/3}{1/2} = \frac{2}{3}.$$

If we switch, there is a $\frac{2}{3}$ chance we will win the prize. If we don't switch, there is only a $\frac{1}{3}$ chance we will win the prize. Thus, switching is better than not switching.

If this reasoning does not appeal to you, there is another way to see that the chance of winning given a switch is $\frac{2}{3}$. In the case of switching, we are making a conscious decision; there is no probabilistic voodoo that is affecting this part of the outcome. So, just consider the outcomes in which we switch and count. Note that there

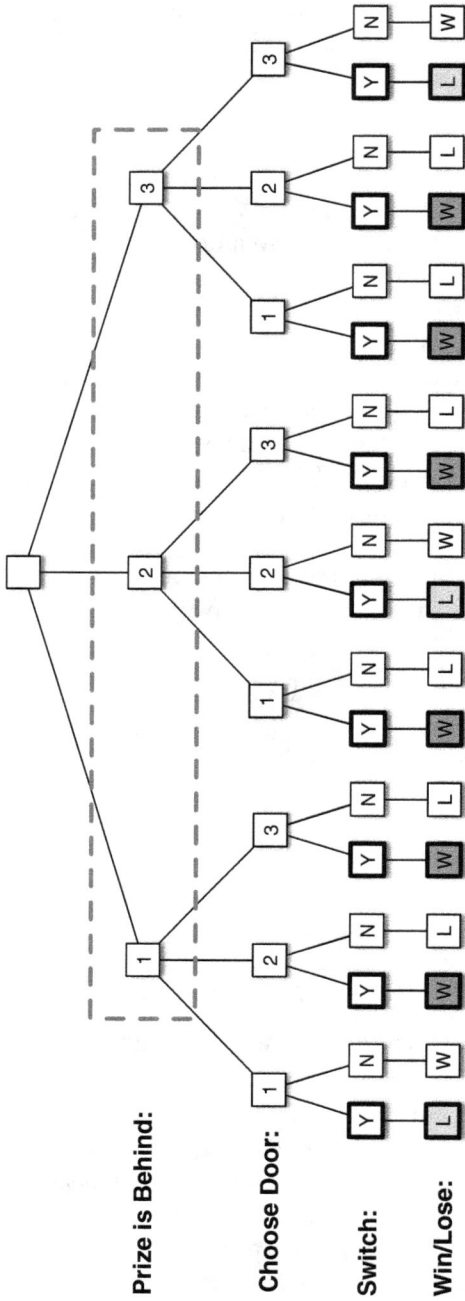

Fig. 1.4. The Monty Hall problem is a multi-stage decision problem whose solution relies on conditional probability. The stages of decision-making are shown in the diagram. We assume that the prizes are randomly assigned to the doors. We cannot see this step, so we adorn this decision with a square box. We discuss these boxes further when we talk about *game trees.* You, the player, must first choose a door. Lastly, you must decide whether to switch doors after being shown an incorrect door.

are 9 outcomes in which we switch from our original door to a door we did not pick first. In 6 of these 9, we win the prize, while in 3, we fail to win the prize. Thus, the chances of winning the prize when we switch is $\frac{6}{9} = \frac{2}{3}$.

1.8   Bayes' Theorem

Remark 1.56. In this final section, we turn to Bayes' theorem and use it to consider the most serious decision problem of all: one related to health. This section is not used in the remainder of the text, but it rounds out the treatment of elementary probability in decision problems. It can be safely skipped with no loss of continuity.

Remark 1.57. In its simplest form, Bayes' theorem can be proved using the definition of conditional probability. Proving the following lemma is left as an exercise.

Lemma 1.58 (Bayes' Theorem – Special Case). *Let $(\Omega, \mathcal{F}, P)$ be a discrete probability space, and suppose that $E, F \in \mathcal{F}$. Then,*

$$P(F|E) = \frac{P(E|F)P(F)}{P(E)}. \tag{1.24}$$

Remark 1.59. We can generalize this result when we have a collection of sets, $F_1, \ldots, F_n \in \mathcal{F}$, that partition Ω and are pairwise disjoint.

Theorem 1.60 (Bayes' Theorem – General Form). *Let $(\Omega, \mathcal{F}, P)$ be a discrete probability space, and suppose that $E, F_1, \ldots, F_n \in \mathcal{F}$, with $F_1, \ldots, F_n$ being pairwise disjoint and*

$$\Omega = \bigcup_{i=1}^{n} F_i.$$

Then,

$$P(F_i|E) = \frac{P(E|F_i)P(F_i)}{\sum_{j=1}^{n} P(E|F_j)P(F_j)}. \tag{1.25}$$

Proof.   Consider the fact that

$$\sum_{j=1}^{n} P(E|F_j)P(F_j) = \sum_{j=1}^{n} \left(\frac{P(E \cap F_j)}{P(F_j)} P(F_j) \right)$$

$$= \sum_{j=1}^{n} P(E \cap F_j) = P(E),$$

by Theorem 1.41. From Lemma 1.58, we conclude that

$$\frac{P(E|F_i)P(F_i)}{\sum_{j=1}^{n} P(E|F_j)P(F_j)} = \frac{P(E|F_i)P(F_i)}{P(E)} = P(F_i|E).$$

This completes the proof.   □

Example 1.61. Here's a rather morbid example: Suppose that a specific disease occurs with a probability of 1 in 1,000,000. A simple test exists to determine whether an individual has this disease. When an individual has the disease, the test will detect it 99 times out of 100. The test also has a false positive rate of 1 in 1,000 (that is, there is a 0.001 probability of misdiagnosis). The treatment for this disease is costly and unpleasant. You have just tested positive. What do you do?

There are two events to consider: (i) the event of having the disease (F) and (ii) the event of testing positive (E). We are interested in computing

$$P(F|E) = \text{The probability of having the disease given a}$$
$$\text{positive test.}$$

We know the following information:

(1) $P(F) = 1 \times 10^{-6}$: There is a 1 in 1,000,000 chance of having this disease.
(2) $P(E|F) = 0.99$: The probability of testing positive given that you have the disease is 0.99.
(3) $P(E|F^c) = 0.001$: The probability of testing positive given that you do not have the disease is 1 in 1,000.

We can apply Bayes' theorem to see that

$$P(F|E) = \frac{P(E|F)P(F)}{P(E|F)P(F) + P(E|F^c)P(F^c)}$$

$$= \frac{(0.99)(1 \times 10^{-6})}{(0.99)(1 \times 10^{-6}) + (0.001)(1 - 1 \times 10^{-6})} = 0.00098.$$

$$(1.26)$$

Thus, the probability of having the disease given the positive test is less than 1 in 1,000. You should follow up with additional tests before committing to the unpleasant and costly treatment.

1.9   Chapter Notes

Modern probability theory began with a letter from the Chevalier de Méreé (pen name of Antoine Gombaud) to Blaise Pascal (of Pascal's triangle fame) [9]. (Though Cardano did work on games of chance well before this [10].) Gombaud was a gambler and wanted to know how to divide winnings in a game of dice when play was interrupted (and the game could not be finished). It is clear from the correspondence that Gombaud kept detailed records of his gambling and was able to furnish Pascal with significant details. Pascal, in turn, corresponded with Pierre de Fermat (of Fermat's last theorem [9] fame), and together they pieced together the essentials of discrete probability distributions. Devlin [11] presents an outstanding history of this interaction.

The work of Pascal and Fermat was extended by Huygens [12], and the classical definition of probability was completed by Laplace [13]. During this time, the reverend Thomas Bayes worked as an amateur statistician and independently published the theorem that now bears his name [14]. Among others, Gauss would later develop additional statistical methods as part of his duties as an astronomer and was among the first to use error analysis in tracking the dwarf planet Ceres using what we now call Gaussian distributions [15], which is why the distribution bears his name. It is worth noting that some authors credit de Moivre with the invention of the normal distribution [16], though this is not widely accepted.

By the late 19th century, it was clear that this treatment of probability would not suffice. David Hilbert challenged the mathematical community to axiomatize probability in the service of statistical physics in his 6th problem. This problem was taken up by Kolmogorov [9], who provided the first measure-theoretic axiomatization of probability theory, which is similar in style to the one presented in this chapter. Kolmogorov was a student of Markov, who contributed to the theory of stochastic processes [9]. For those interested in an introduction to measure theory, Bear [17] has a short and readable book on the subject.

Card counting is almost an immediate consequence of Gambaoud's letter to Pascal, though it was the mathematician Edward O. Thorp who proved that card counting could be used to overcome house advantage in blackjack in his book *Beat the Dealer* [18]. Thorp's work was highly original and influential, but was influenced by the prior work of J. L. Kelly of Bell Labs, who established the *Kelly criterion* for formulating bet size [19].

The chief roadblock to card counters is knowing the count before sitting at the table. The MIT card-counting team (featured in the movie *21*) used a *big player team* strategy [20]. In this strategy, card counters would sit at a table and make safe bets, winning or losing very little over the course of time. They would keep the card count and signal *big players* from their team, who would arrive at the table and make large bets when the count was high (in their favor). The big players would leave once signaled that the count had dropped. Using this strategy, the MIT players cleared millions from the casinos using basic probability theory.

In response to the academic work on card counting, casinos have invested time and money in detection methods. Detecting card counting has become a major part of casino *intelligence operations*. There are several extremely sophisticated methods casinos employ for detecting counting strategies. In general, it is difficult to make counting work in a modern casino without using a team system, and careful hedging is required to stay below the radar of the pit boss and the eye in the sky. We note that the gambling example given in this chapter is of a very unsophisticated version of blackjack. Most casinos use six decks at a blackjack table, introduce fresh packs of cards, and employ automatic shufflers designed to negatively impact counting. When all else fails, dealers may intentionally distract players they suspect are counters to break their concentration and ruin the count.

The *Monty Hall problem* discussed in this chapter first appeared in 1975 in the *American Statistician* by Steve Selvin [21]. It was subsequently described by Morgan, Chaganty, Dahiya, and Doviak [22], again in the *American Statistician*, who pointed out that the assumptions in the Monty Hall problem setup are critical to the solution. The decision problem was then discussed in follow-up letters and articles [23–25], and in 2002, a "quantum" version was formulated and analyzed [26]. A historical account of the problem and its solutions was written by Rosenhouse in 2009 [27]. If the problem is rephrased as a two-player game in which the host stands to gain, it can radically alter the solution structure. The problem then becomes one of determining problem setups in which switching is always better. Interestingly, in a 1991 interview [28], Monty Hall revealed that he had control over which deals to offer and was therefore able to manipulate the contestants psychologically and affect their decisions. Thus, in real life, *Let's Make a Deal* was more similar to a two-player game than a game against the house.

$$- \spadesuit \clubsuit \heartsuit \diamondsuit -$$

1.10   Exercises

1.1   A fair four-sided die is rolled. Assume that the sample space of interest is the number appearing on the dice, and the numbers run from 1 to 4. Identify the space Ω precisely and all the possible outcomes and events within the space. What is the (logical) fair probability distribution in this case. [Hint: See Example 1.18.]

1.2   Compute the probability of rolling a double 6 in 24 rolls of a pair of dice. [Hint: Each roll is independent of the last roll. Let E be the event that you *do not* roll a double 6 on a given roll. The probability of E is $35/36$ (i.e., there are 35 other ways the dice could come out other than double 6). Now, compute the probability of not seeing a double six in all 24 rolls using independence. (You will get a power of 24.) Let this probability be p. Finally, note that the probability of a double 6 occurring is precisely $1 - p$. To see this, note that p is the probability of the event that a double six does not occur. Thus, the probability of the event that a double 6 does occur must be $1 - p$.]

1.3   Prove the following: Let $E \subseteq \Omega$ and define E^c to be the set of elements of Ω *not* in E (which is called the complement of E). Suppose $(\Omega, \mathcal{F}, P)$ is a discrete probability space. Show that $P(E^c) = 1 - P(E)$.

1.4   Prove Lemma 1.38. [Hint: Show that $E \cap F$ and $E \cap F^c$ are mutually exclusive events. Then, show that $E = (E \cap F) \cup (E \cap F^c)$.]

1.5   Use Definition 1.47 to compute the probability of obtaining a sum of 8 in two rolls of a die, given that in the first roll, a 1 or 2 appears. [Hint: The space of outcomes is still $\Omega = \{(x, y) | x = 1, \ldots, 6, \ y = 1, \ldots, 6\}$. First, identify the event E within this space. How many elements within this set will enable you to obtain an 8 in two rolls? This is the set $E \cap F$. What is the probability of $E \cap F$? What is the probability of E? Use the formula in Definition 1.47. It might help to write out the space Ω.]

1.6   Show (in any way you like) that the probability of winning in the Monty Hall problem given that you *do not* switch doors, is $1/3$.

1.7   In the little-known *Lost Episodes of Let's Make a Deal*, Monty (or Wayne) introduces a fourth door. Suppose that you choose a door and then you are shown two incorrect doors and given the chance to switch. Should you switch? Why? [Hint: Build a figure similar to Fig. 1.4. It will be a bit large. Use the same reasoning we used to compute the probability of successfully winning the prize in the previous example.]

1.8   Prove the simple form of Bayes' theorem. [Hint: Use Definition 1.47.]

1.9   In Example 1.61, for what probability of having the disease is there a 1 in 100 chance of having the disease, given that you have tested positive? [Hint: I'm asking for what value of $P(F)$ is the value of $P(F|E)$ 1 in 100. Draw a graph of $P(F|E)$ and use a calculator.]

1.10   There are several other ways to analyze the Monty Hall problem. Use Bayes' theorem to show that the probability of winning in the Monty Hall problem is $2/3$.

Chapter 2

Elementary Utility Theory

Chapter Goals: The goal of this chapter is to introduce utility theory, as defined by von Neumann and Morgenstern [5], and to use it to show how even non-monetary prizes can be transformed into numerical payoffs. We also introduce the expectation maximization theorem. This chapter can be skipped by those readers who want to get into multiplayer game theory, as long as you accept the central premise that all prizes can be converted into numeric payoffs under certain conditions.

2.1 Decision-Making Under Certainty

Remark 2.1. The game show *The Price is Right* premiered in the United States in 1972 and features contestants competing in various games against the house to win prizes and occasionally money. Almost all games involve an element of chance and knowledge about the prices of commercial products [29]. This represents a stark contrast from *Deal or No Deal*, where the prizes are all monetary in nature.

Definition 2.2 (Lottery). A *lottery*, $L = \langle \{A_1, \ldots, A_n\}, P \rangle$, is a collection of prizes (or rewards, or costs), $\{A_1, \ldots, A_n\}$, along with a discrete probability distribution, P, with the sample space $\{A_1, \ldots, A_n\}$. We denote the set of all lotteries over $A_1, \ldots, A_n$ by $\mathcal{L}$.

Remark 2.3. In a lottery (of this type), we do not assume that we will determine the probability distribution P as a result of repeated exposure. (This is not like a state lottery.) Instead, the probability is given *ab initio* and does not change. Also, *ab initio* is a fancy way of saying "from the beginning."

Remark 2.4. To simplify notation, we state that

$$L = \langle (A_1, p_1), \ldots, (A_n, p_n) \rangle$$

is the lottery consisting of prizes A_1–A_n, where you receive prize A_1 with a probability of p_1, prize A_2 with a probability of p_2, etc. Note that it is assumed in the definition that

$$p_1 + p_2 + \cdots + p_n = 1$$

because these probabilities arise from a discrete probability function with sample space $A_1, \ldots, A_n$.

Remark 2.5. The lottery in which we win the prize A_i with a probability of 1 and all other prizes with a probability of 0 will be denoted as A_i as well. Thus, the prize A_i is equivalent to a lottery in which one always wins the prize A_i.

Example 2.6. On *The Price is Right*, in the game *Temptation*, the contestant is offered four (small) prizes and given their dollar value. From the dollar values, the contestant must then construct the price of a large prize (e.g., a car). Once the player is shown all the prizes (and constructs a guess for the price of the car), the player must make a choice between taking the small prizes and leaving or risking the prizes and playing for the large prize.

In this example, there are two lotteries: the small prize option and the large prize option. The small prize option contains a single reward consisting of the various items seen by the contestant. Denote this lottery as A_1. This lottery is (A_1, P_1), where $P_1(A_1) = 1$. The other lottery option contains two rewards: the large prize (car) A_2 and the null prize A_0 (where the contestant leaves with nothing). This lottery has the form $\langle \{A_0, A_2\}, P_2 \rangle$, where $P_2(A_0) = p$, $P_2(A_2) = 1 - p$, and $p \in (0, 1)$, and depends on the nature of the prices of the prizes in A_1, which were used to construct the guess for the price of the large prize.

2.2 Preference and the von Neumann–Morgenstern Assumptions

Definition 2.7 (Preference). Let L_1 and L_2 be lotteries. We write $L_1 \succeq L_2$ to indicate that an individual *prefers* lottery L_1 to lottery L_2. If both $L_1 \succeq L_2$ and $L_2 \succeq L_1$, then $L_1 \sim L_2$, and L_1 and L_2 are considered equivalent to the individual.

Remark 2.8. The axiomatic treatment of utility theory rests on certain assumptions about an individual's behavior when they are confronted with a choice of two or more lotteries. We have already seen this type of scenario in Example 2.6. We assume that these choices are governed by preference. Preferences can vary from individual to individual.

Remark 2.9. For the remainder of this chapter, we assume that every lottery consists of prizes $A_1, \ldots, A_n$ and that the prizes are preferred in the order

$$A_1 \succeq A_2 \succeq \cdots \succeq A_n. \tag{2.1}$$

We now introduce five assumptions that will be needed to prove the *expected utility theorem*.

Assumption 1. Let L_1, L_2 and L_3 be lotteries:

(1) Either $L_1 \succeq L_2$ or $L_2 \succeq L_1$ or $L_1 \sim L_2$.
(2) If $L_1 \preceq L_2$ and $L_2 \preceq L_3$, then $L_1 \preceq L_3$.
(3) If $L_1 \sim L_2$ and $L_2 \sim L_3$, then $L_1 \sim L_3$.
(4) If $L_1 \succeq L_2$ and $L_2 \succeq L_1$, then $L_1 \sim L_2$.

Remark 2.10. Item 1 of Assumption 1 states that the ordering $\preceq$ is a total ordering on the set of all lotteries with which an individual may be presented. That is, we can compare any two lotteries to each other and always be able to decide which one is preferred or whether they are equivalent. Item 2 of Assumption 1 states that this ordering is *transitive*.

It is clear that preference should be *reflexive* (i.e., $L_1 \sim L_1$ for all lotteries L_1) and *symmetric* ($L_1 \sim L_2$ if and only if $L_2 \sim L_1$ for all lotteries L_1 and L_2). The assumption of transitivity implies that preferential equivalence is an *equivalence relation* over the set of all lotteries.

Example 2.11 (Is Transitivity a Problem?). For this example, you must use your imagination and think like a preschooler. Imagine a scenario in which we present a preschooler with the following choices (lotteries with only one item): a ball, a puzzle, and a crayon (and paper). If we present the choice of the puzzle and the crayon, the child may choose the crayon. In presenting the puzzle and the ball, the child may choose the puzzle. On the other hand, suppose we present the crayon and the ball. If the child chooses the ball, then transitivity is violated. We can ask why this might happen. It is possible that the child's preferences will change depending upon the current requirements of their imagination. Such transitivity violations have been observed in real-world experiments [30].

Definition 2.12 (Compound Lottery). Let $L_1, \ldots, L_n$ be a set of lotteries, and suppose that the probability of being presented with lottery i ($i = 1, \ldots, n$) is q_i. A lottery $Q = \langle (L_1, q_1), \ldots, (L_n, q_n) \rangle$ is called a *compound lottery*.

Example 2.13. Consider a fictional game called *Flip of a Coin!*. Two contestants are randomly assigned heads or tails. A coin is flipped:

- The winner is offered a chance to flip a second coin or to leave with a guaranteed \$500. If the second coin comes up heads, the winner receives \$1,000; if the coin comes up tails, the contestant flips an unfair coin that comes up heads 10% of the time. If the contestant flips heads, she wins \$10,000. If the contestant flips tails, she leaves with nothing.
- The loser is offered the choice of leaving with nothing or flipping a second coin. If the second coin comes up heads, the loser receives \$100. If the coin comes up tails, the contestant flips an unfair coin that comes up heads 10% of the time. If the contestant flips heads, he wins \$1,000. If the contestant flips tails, he falls into a tank of water and leaves with nothing but wet clothes.

We can work backward to model the entire show as a collection of (compound) lotteries. Unless stated otherwise, assume all coin flips are fair and that the loser opts to flip the second coin. Then, there are two possible lotteries he will see:

$$L_T = \langle (\$1{,}000, \tfrac{1}{10}), (\text{Wet}, \tfrac{9}{10}) \rangle.$$
$$L_H = \langle (\$100, 1) \rangle.$$

The lottery L_T occurs if the second coin flip comes up tails. The lottery L_H occurs if the second coin flip comes up heads. Note that this is a trivial lottery with a guaranteed prize.

The losing contestant must choose between a trivial lottery,

$$L_1 = \langle (0,1) \rangle,$$

in which he gets nothing, and a compound lottery,

$$L_2 = \langle \left(L_H, \tfrac{1}{2} \right), \left(L_T, \tfrac{1}{2} \right) \rangle.$$

For the winning contestant, if she opts to flip the second coin, the outcomes can be again modeled with two lotteries:

$$W_T = \langle \left(\$10,000, \tfrac{1}{10} \right), \left(\$0, \tfrac{9}{10} \right) \rangle$$

and

$$W_H = \langle (\$1,000, 1) \rangle.$$

Since these are again decided by a coin flip, the compound lottery in the winning contestant's choice is

$$W_2 = \langle \left(W_H, \tfrac{1}{2} \right), \left(W_T, \tfrac{1}{2} \right) \rangle.$$

The winning contestant chooses between this compound lottery and the trivial lottery

$$W_1 = \langle (\$500, 1) \rangle.$$

The possible lotteries and outcomes (with decisions) are illustrated in Fig. 2.1.

Assumption 2. Let $\langle (L_1, q_1), \ldots, (L_n, q_n) \rangle$ be a compound lottery, and suppose each L_i $(i = 1, \ldots, n)$ is composed of prizes $A_1, \ldots, A_m$ with probabilities p_{ij} $(j = 1, \ldots, m)$. Then, this compound lottery is equivalent to a simple lottery in which the probability of prize A_j is

$$r_j = q_1 p_{1j} + q_2 p_{2j} + \cdots + q_n p_{nj}.$$

Remark 2.14. Assumption 2 states that compound lotteries can be transformed into equivalent simple lotteries. Note further that the probability of prize j (A_j) is actually

$$P(A_j) = \sum_{i=1}^{n} P(A_j | L_i) P(L_i). \tag{2.2}$$

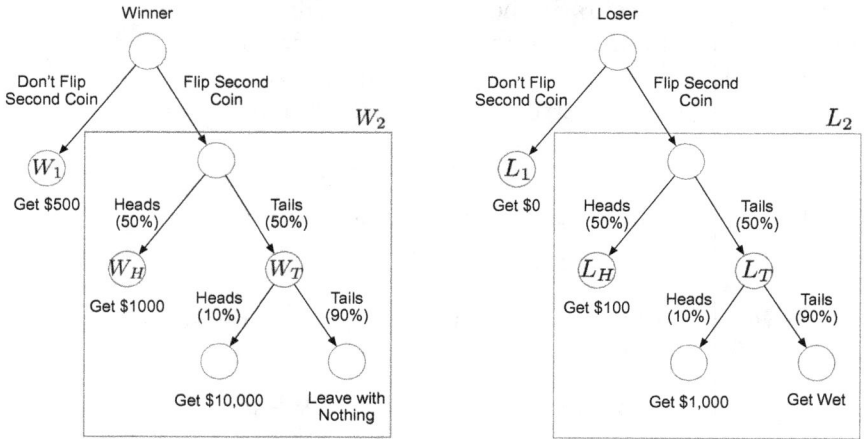

Fig. 2.1. The possible outcomes for "Flip a Coin!" are illustrated. Simple lotteries are labeled in the circles in which a random event occurs. Compound lotteries are shown with the boxes.

This statement should be clear from Theorem 1.41, when we define a probability space in the correct manner.

Example 2.15. Consider the winner's compound lottery from Example 2.13. Recall that the lotteries were

$$W_2 = \left\langle \left(W_H, \tfrac{1}{2}\right), \left(W_T, \tfrac{1}{2}\right) \right\rangle,$$

$$W_T = \left\langle \left(\$10{,}000, \tfrac{1}{10}\right), \left(\$0, \tfrac{9}{10}\right) \right\rangle,$$

$$W_H = \left\langle (\$1{,}000, 1) \right\rangle.$$

The prizes in this case are \$10,000, \$1,000, and \$0. Then, we can compute the probabilities associated with the different prizes:

$$p_{\$0} = \underbrace{\frac{1}{2} \cdot \frac{9}{10}}_{\text{From } W_T} + \underbrace{\frac{1}{2} \cdot 0}_{\text{From } W_H} = \frac{9}{20},$$

$$p_{\$1,000} = \underbrace{\frac{1}{2} \cdot 0}_{\text{From } W_T} + \underbrace{\frac{1}{2} \cdot 1}_{\text{From } W_H} = \frac{1}{2},$$

$$p_{\$10,000} = \underbrace{\frac{1}{2} \cdot \frac{1}{10}}_{\text{From } W_T} + \underbrace{\frac{1}{2} \cdot 0}_{\text{From } W_H} = \frac{1}{20}.$$

Since these prizes are all monetary, we can compute the expected payoff to the winner of the first prize using the simple lottery representation. If W is the random variable (not the lottery) giving the payoff to the winner, assuming she flips the second coin, the payoff is

$$\mathbb{E}(W) = 0 \cdot \frac{9}{20} + 1000 \cdot \frac{1}{2} + 10000 \cdot \frac{1}{20} = 1000.$$

Remark 2.16. In Example 2.15, the winner receives only monetary prizes; therefore, it is easy to compare her possible decisions. In contrast, it is possible the loser will land in a tank of water, which does not have an immediate monetary value. The remaining assumptions will allow us to construct a mapping from lotteries to numeric values, which can then be compared directly so that we do not have to convert cars to dollars or understand the monetary loss associated with wet clothes.

Assumption 3. For each prize (or lottery) A_i, there is a number $u_i \in [0, 1]$ so that the prize A_i (or lottery L_i) is preferentially equivalent to the lottery in which you win prize A_1 with a probability of u_i, A_n with a probability of $1 - u_i$, and all other prizes with a probability of 0. This lottery will be denoted $\tilde{A}_i$.

Remark 2.17. Assumption 3 is often called the *continuity* assumption, and it is a little strange. It assumes that for any ordered set of prizes $(A_1, \ldots, A_n)$, a person would view winning any specific prize (A_i) as equivalent to playing a game of chance in which either the worst or best prize could be obtained.

This assumption may not be valid in all cases. Suppose that the best prize A_1 is a new car, while the worst prize A_n is spending 10 years in jail. If the prize in question (A_i) is that you receive \$100, then the continuity assumption implies that there is a game of chance you would play involving a new car or 10 years in jail that would be equal to receiving \$100.

Assumption 4. If $L = \langle (A_1, p_1), \ldots, (A_i, p_i), \ldots, (A_n, p_n) \rangle$ is a lottery, then L is preferentially equivalent to the lottery $\langle (A_1, p_1), \ldots, (\tilde{A}_i, p_i), \ldots, (A_n, p_n) \rangle$.

Remark 2.18. Assumption 4 only asserts that we can substitute any equivalent lottery for a prize and not change the preferential ordering.

Assumption 5. A lottery L in which A_1 is obtained with a probability of p and A_n is obtained with a probability of $(1 - p)$ is always preferred or equivalent to a lottery in which A_1 is obtained with a probability of p' and A_n is obtained with a probability of $(1 - p')$ if and only if $p \geq p'$.

Remark 2.19. Assumption 5 states that anyone would prefer (or at least be indifferent to) winning A_1 with a higher probability and A_n with a lower probability. This assumption is reasonable when we have the case $A_1 \succeq A_n$. In Ref. [1], it is pointed out that there are psychological reasons why this assumption may be violated.

2.3   Expected Utility Theorem

Remark 2.20. The *expected utility theorem* is the result of all these assumptions. It provides a formal way of assigning numerical values to prizes, even if those prizes have no obvious numerical value. Its proof rests on the five assumptions. If any of those assumptions are violated, then this theorem need not be true.

Theorem 2.21 (Expected Utility Theorem). *Let $\succeq$ be a preference relation satisfying Assumptions 1–5 over the set of all lotteries $\mathcal{L}$ defined over the prizes $A_1, \ldots, A_n$. Furthermore, assume that*

$$A_1 \succeq A_2 \succeq \cdots \succeq A_n.$$

Then, there is a function $u : \mathcal{L} \to [0, 1]$ with the property that

$$u(L_1) \geq u(L_2) \iff L_1 \succeq L_2. \tag{2.3}$$

Proof.   We begin by defining the utility function:

(1) Define $u(A_1) = 1$. Recall that A_1 is not only prize A_1 but also the lottery in which we receive A_1 with a probability of 1, that is, the lottery in which $p_1 = 1$ and $p_2 \ldots, p_n = 0$.
(2) Define $u(A_n) = 0$. Again, recall that A_n is also the lottery in which we receive A_n with a probability of 1.
(3) By Assumption 3, for lottery A_i ($i \neq 1$ and $i \neq n$), there is a u_i so that A_i is equivalent to $\tilde{A}_i$: the lottery in which you win prize A_1 with a probability of u_i, A_n with a probability of $1 - u_i$, and all other prizes with a probability of 0. Define $u(A_i) = u_i$.

(4) Let $L \in \mathcal{L}$ be a lottery in which we win prize A_i with a probability of p_i. Then,

$$u(L) = p_1 u_1 + p_2 u_2 + \cdots + p_n u_n. \tag{2.4}$$

Here, $u_1 \equiv 1$ and $u_n \equiv 0$.

We now show that this utility function satisfies Eq. (2.3).

($\Leftarrow$) Let $L_1, L_2 \in \mathcal{L}$, and suppose that $L_1 \succeq L_2$. Suppose that

$$L_1 = \langle (A_1, p_1), (A_2, p_2), \ldots, (A_n, p_n) \rangle,$$
$$L_2 = \langle (A_1, q_1), (A_2, q_2), \ldots, (A_n, q_n) \rangle.$$

By Assumption 3, for each A_i, $(i \neq 1, i \neq n)$, we know that $A_i \sim \tilde{A}_i$, with $\tilde{A}_i \equiv \langle (A_1, u_i), (A_n, 1 - u_i) \rangle$. Then, by Assumption 4, we know that

$$L_1 \sim \langle (A_1, p_1), (\tilde{A}_2, p_2), \ldots, (\tilde{A}_{n-1}, p_{n-1}), (A_n, p_n) \rangle,$$
$$L_2 \sim \langle (A_1, q_1), (\tilde{A}_2, q_2), \ldots, (\tilde{A}_{n-1}, q_{n-1}), (A_n, q_n) \rangle.$$

These are compound lotteries, and we can expand them as

$$L_1 \sim \langle (A_1, p_1), (\langle (A_1, u_2), (A_n, (1 - u_2)) \rangle, p_2), \ldots,$$
$$(\langle (A_1, u_{n-1}), (A_n, (1 - u_{n-1})) \rangle, p_{n-1}), (A_n, p_n) \rangle. \tag{2.5}$$
$$L_2 \sim \langle (A_1, q_1), (\langle (A_1, u_2), (A_n, (1 - u_2)) \rangle, q_2), \ldots,$$
$$(\langle (A_1, u_{n-1}), (A_n, (1 - u_{n-1})) \rangle, q_{n-1}), (A_n, q_n) \rangle. \tag{2.6}$$

We can apply Assumption 2 to transform these compound lotteries into simple lotteries. By combining the like prizes, we see that

$$L_1 \sim \langle (A_1, p_1 + u_2 p_2 + \cdots + u_{n-1} p_{n-1}),$$
$$(A_n, (1 - u_2) p_2 + \cdots + (1 - u_{n-1}) p_{n-1} + p_n) \rangle \stackrel{\Delta}{=} \tilde{L}_1,$$

and

$$L_2 \sim \langle (A_1, q_1 + u_2 q_2 + \cdots + u_{n-1} q_{n-1}),$$
$$(A_n, (1 - u_2) q_2 + \cdots + (1 - u_{n-1}) q_{n-1} + q_n) \rangle \stackrel{\Delta}{=} \tilde{L}_2.$$

Here, $\stackrel{\Delta}{=}$ simply means that we define $\tilde{L}_1$ and $\tilde{L}_2$ by these expressions. We can apply Assumption 1 to see that $L_1 \sim \tilde{L}_1$, $L_2 \sim \tilde{L}_2$, and

$L_1 \succeq L_2$. This implies that $\tilde{L}_1 \succeq \tilde{L}_2$. By Assumption 5, we conclude that

$$p_1 + u_2 p_2 + \cdots + u_{n-1} p_{n-1} \geq q_1 + u_2 q_2 + \cdots + u_{n-1} q_{n-1}. \quad (2.7)$$

Note, however, that

$$u(L_1) = p_1 + u_2 p_2 + \cdots + u_{n-1} p_{n-1},$$
$$u(L_2) = q_1 + u_2 q_2 + \cdots + u_{n-1} q_{n-1}.$$

Thus, we have $u(L_1) \geq u(L_2)$.

$(\Rightarrow)$ Suppose now that $L_1, L_2 \in \mathcal{L}$ and that $u(L_1) \geq u(L_2)$. Then we know that,

$$u(L_1) = u_1 p_1 + u_2 p_2 + \cdots + u_{n-1} p_{n-1} + u_n p_n$$
$$\geq u_1 q_1 + u_2 q_2 + \cdots + u_{n-1} q_{n-1} + u_n q_n = u(L_2). \quad (2.8)$$

As before, we have $L_1 \sim \tilde{L}_1$ and $L_2 \sim \tilde{L}_2$. We also have $u(L_1) = u(\tilde{L}_1)$ and $u(L_2) = u(\tilde{L}_2)$. To see this, note that in $\tilde{L}_1$, the probability associated with prize A_1 is

$$p_1 + u_2 p_2 + \cdots + u_{n-1} p_{n-1}.$$

Thus (since $u_1 = 1$ and $u_n = 0$), we know that

$$u(\tilde{L}_1) = u_1 \left(p_1 + u_2 p_2 + \cdots + u_{n-1} p_{n-1} \right)$$
$$= u_1 p_1 + u_2 p_2 + \cdots + u_{n-1} p_{n-1} + u_n p_n.$$

A similar statement holds for $\tilde{L}_2$, and thus we can conclude that

$$p_1 + u_2 p_2 + \cdots + u_{n-1} p_{n-1} \geq q_1 + u_2 q_2 + \cdots + u_{n-1} q_{n-1}. \quad (2.9)$$

We can now apply Assumption 5 (which is an if-and-only-if statement) to see that

$$\tilde{L}_1 \succeq \tilde{L}_2.$$

We can now conclude from Assumption 1 that, since $L_1 \sim \tilde{L}_1$, $L_2 \sim \tilde{L}_2$, and $\tilde{L}_1 \succeq \tilde{L}_2$, $L_1 \succeq L_2$. This completes the proof. $\qquad \square$

Remark 2.22. This theorem is called the expected utility theorem because the utility of any lottery is in fact the expected utility of any of the prizes. That is, let U be the random variable that takes value u_i if prize A_i is received. Then,

$$\mathbb{E}(U) = \sum_{i=1}^{n} u_i p(A_i) = u_1 p_1 + u_2 p_2 + \cdots + u_n p_n. \qquad (2.10)$$

This is simply the utility of the lottery in which prize i is received with a probability of p_i.

Example 2.23. Suppose you are a contestant on *Let's Make a Deal*, and the following prizes are available:

(1) A_1: A new car (worth \$20,000).
(2) A_2: A gift card (worth \$1,000).
(3) A_3: A new iPad (worth \$800).
(4) A_4: A donkey (technically worth \$500, but somewhat challenging).

We assume that you prefer these prizes in the order in which they appear. The host offers you the following deal: You can compete in either of the following games (lotteries):

(1) $L_1 = \langle (A_1, 0.25), (A_2, 0.25), (A_3, 0.25), (A_4, 0.25) \rangle$;
(2) $L_2 = \langle (A_1, 0.15), (A_2, 0.4), (A_3, 0.4), (A_4, 0.05) \rangle$.

Which games should you choose to make you the most happy? The problem here is valuing the prizes. Maybe you really need a new car (or you just bought a new car). The car may be worth more than its dollar value. Alternatively, suppose you actually want a donkey. Suppose you know that donkeys are expensive to own and the "retail" value (\$500) is false. Perhaps it would be difficult to use a gift card of that size.

For the sake of argument, let's suppose that you determine that the donkey is worth nothing to you. You might say that:

(1) $A_2 \sim \langle (A_1, 0.1), (A_4, 0.9) \rangle$;
(2) $A_3 \sim \langle (A_1, 0.05), (A_4, 0.95) \rangle$.

This implies that $u(A_2) = u_2 = 0.1$ and $u(A_3) = u_3 = 0.05$. The numbers really don't make any difference; you can supply any values you want for 0.1 and 0.05 as long as the other numbers enforce

Assumption 3. Then, we can write

$$L_1 \sim \langle (A_1, 0.25), (\langle\langle(A_1, 0.1), (A_4, 0.9)\rangle\rangle, 0.25),$$
$$(\langle\langle(A_1, 0.05), (A_4, 0.95)\rangle\rangle, 0.25), (A_4, 0.25)\rangle,$$
$$L_2 \sim \langle (A_1, 0.15), (\langle\langle(A_1, 0.1), (A_4, 0.9)\rangle\rangle, 0.4),$$
$$(\langle\langle(A_1, 0.05), (A_4, 0.95)\rangle\rangle, 0.4), (A_4, 0.05)\rangle.$$

We can now simplify this by expanding these compound lotteries into simple lotteries in terms of A_1 and A_4. To see how we do this, consider only Lottery 1. Lottery 1 is a compound lottery that contains the following sub-lotteries:

(1) S_1: A_1 with a probability of 0.25;
(2) S_2: $\langle(A_1, 0.1), (A_4, 0.9)\rangle$ with a probability of 0.25;
(3) S_3: $\langle(A_1, 0.05), (A_4, 0.95)\rangle$ with a probability of 0.25;
(4) S_4: A_4 with a probability of 0.25.

To convert this lottery into a simpler lottery, apply Assumption 2. We have

$$P(A_1) = \underbrace{P(A_1|S_1)}_{1} \cdot \underbrace{P(S_1)}_{0.25} + \underbrace{P(A_1|S_2)}_{0.1} \cdot \underbrace{P(S_2)}_{0.25}$$
$$+ \underbrace{P(A_1|S_3)}_{0.05} \cdot \underbrace{P(S_3)}_{0.25} + \underbrace{P(A_1|S_4)}_{0} \cdot \underbrace{P(S_4)}_{0.25} = 0.2875.$$

Using a similar computation, we deduce that

$$P(A_4) = 0.71250.$$

We can conclude that

$$L_1 \sim \langle (A_1, 0.2875), (A_4, 0.71250) \rangle \triangleq \tilde{L}_1.$$

We can perform a similar calculation for L_2 to get

$$L_2 \sim \langle (A_1, 0.21), (A_4, 0.79) \rangle \triangleq \tilde{L}_2.$$

It is now easy to compute the utility of the two games. We know immediately that

$$u(L_1) = u(\tilde{L}_1) = 0.2875,$$
$$u(L_2) = u(\tilde{L}_2) = 0.21.$$

Computing $u(L_1)$ and $u(L_2)$ using Eq. (2.4) would yield the same result:

$$u(L_1) = \underbrace{p_1}_{0.25} \underbrace{u_1}_{1} + \underbrace{p_2}_{0.25} \underbrace{u_2}_{0.1} + \underbrace{p_3}_{0.25} \underbrace{u_3}_{0.05} + \underbrace{p_4}_{0.25} \underbrace{u_4}_{0} = 0.2875,$$

$$u(L_2) = \underbrace{p_1}_{0.15} \underbrace{u_1}_{1} + \underbrace{p_2}_{0.4} \underbrace{u_2}_{0.1} + \underbrace{p_3}_{0.4} \underbrace{u_3}_{0.05} + \underbrace{p_4}_{0.05} \underbrace{u_4}_{0} = 0.21.$$

Thus, you should opt to play the first game (L_1), even though there is a greater chance of winning the donkey because L_1 has a higher expected utility than L_2.

Definition 2.24 (Linear Utility Function). We say that a utility function $u : \mathcal{L} \to \mathbb{R}$ is *linear* if, given any lotteries $L_1, L_2 \in \mathcal{L}$ and some $q \in [0, 1]$, then

$$u\left[\langle(L_1, q), (L_2, (1 - q))\rangle\right] = qu(L_1) + (1 - q)u(L_2). \tag{2.11}$$

Here, $\langle(L_1, q), (L_2, (1 - q))\rangle$ is the compound lottery made up of the lotteries L_1 and L_2, each having probabilities of q and $(1-q)$, respectively.

Lemma 2.25. *Let $\mathcal{L}$ be the collection of lotteries defined over prizes $A_1, \ldots, A_n$, with $A_1 \succeq A_2 \succeq \cdots \succeq A_n$. Let $u : \mathcal{L} \to [0, 1]$ be the utility function defined in Theorem 2.21. Then, $L_1 \sim L_2$ if and only if $u(L_1) = u(L_2)$.*

Theorem 2.26. *The utility function $u : \mathcal{L} \to [0, 1]$ defined in Theorem 2.21 is linear.*

Proof. Let

$$L_1 = \langle(A_1, p_1), (A_2, p_2), \ldots, (A_n, p_n)\rangle,$$
$$L_2 = \langle(A_1, r_1), (A_2, r_2), \ldots, (A_n, r_n)\rangle.$$

Thus, we know that

$$u(L_1) = \sum_{i=1}^{n} p_i u_i,$$

$$u(L_2) = \sum_{i=1}^{n} r_i u_i.$$

Choose $q \in [0,1]$. The lottery $L = \langle (L_1, q), (L_2, (1-q)) \rangle$ is equivalent to a lottery in which prize A_i is obtained with a probability of

$$\Pr(A_i) = qp_i + (1-q)r_i.$$

Thus, applying Assumption 2, we have

$$\tilde{L} \triangleq \langle (A_1, [qp_1 + (1-q)r_1]), \ldots, (A_n, [qp_n + (1-q)r_n]) \rangle \sim L.$$

Applying Lemma 2.25, we can compute

$$u(L) = u(\tilde{L}) = \sum_{i=1}^{n} [qp_i + (1-q)r_i] u_i = \sum_{i=1}^{n} qp_i u_i + \sum_{i=1}^{n} (1-q)r_i u_i$$

$$= q \left(\sum_{i=1}^{n} p_i u_i \right) + (1-q) \left(\sum_{i=1}^{n} r_i u_i \right) = qu(L_1) = (1-q)u(L_2).$$

$$(2.12)$$

Thus, u is linear. $\qquad\square$

Theorem 2.27. *Suppose that $a, b \in \mathbb{R}$ with $a > 0$. Then, the function $u' : \mathcal{L} \to \mathbb{R}$, given by*

$$u'(L) = au(L) + b, \qquad (2.13)$$

also has the property that $u'(L_1) \geq u'(L_2)$ if and only if $L_1 \succeq L_2$, where u is the utility function given in Theorem 2.21. Furthermore, this utility function is linear.

Remark 2.28. A generalization of Theorem 2.27 simply shows that the class of linear utility functions is closed under a subset of affine transforms. That means, given a linear utility function, we can construct another by multiplying by a positive constant and adding another constant.

2.4 Chapter Notes

The expected utility theorem and the theory of lotteries, as presented in this chapter, are due to von Neumann and Morgenstern [5] and described in their book *Theory of Games and Economic Behavior*. Oskar Morgenstern was a German economist and is the cofounder of modern game theory, along with von Neumann. Von Neumann, a child prodigy, made seminal contributions to both pure and applied mathematics, as well as physics and computer science [31]. As two examples, the von Neumann architecture is the *de facto* standard computer architecture now used, and von Neumann algebras are named in his honor. Von Neumann was famously witty. Freeman Dyson recalled Enrico Fermi quoting von Neumann as saying, "[W]ith four parameters I can fit an elephant, and with five I can make him wiggle his trunk." [32]. This is in reference to the ability of models with multiple parameters to fit arbitrarily complex datasets without providing any additional insight. Fitting an "elephant" became a problem in recreational mathematics, with the current best "fit" being the parametric equations,

$$x = 30\sin(t) - 8\sin(2t) + 10\sin(3t) - 60\cos(t)$$

$$y = 50\sin(t) + 18\sin(2t) - 12\cos(3t) + 14\cos(5t),$$

for $t \in [0, 2\pi]$. See Fig. 2.2.

This was discovered by Mayer, Khairy, and Howard [33] in 2010 using Fourier analysis with four parameters. The result was then converted into the parametric equations given above. A fifth parameter can be used to make the elephant's eye.

A counterexample to the continuity assumption in Remark 2.17 is a variation of Bernoulli's St. Petersburg paradox, which posits a lottery with an infinite payoff and asks how much anyone would be willing to risk on such a (non-guaranteed) payoff [34]. From this, Bernoulli concluded that taking only the expected reward into account was not rational. Consequently, acceptance of von Neumann and Morgenstern's axioms implies rejection of Bernoulli's conclusion (and vice versa).

There are several variations of utility theory. The one presented in this chapter is usually called the von Neumann–Morgenstern utility. Other theories of utility may assume determinism. Fishburn [35] has a survey of classical utility theory from the perspective of management science, which is valid up to 1968. A more recent survey by Karni and Schmeidler [36] provides information on utility theory

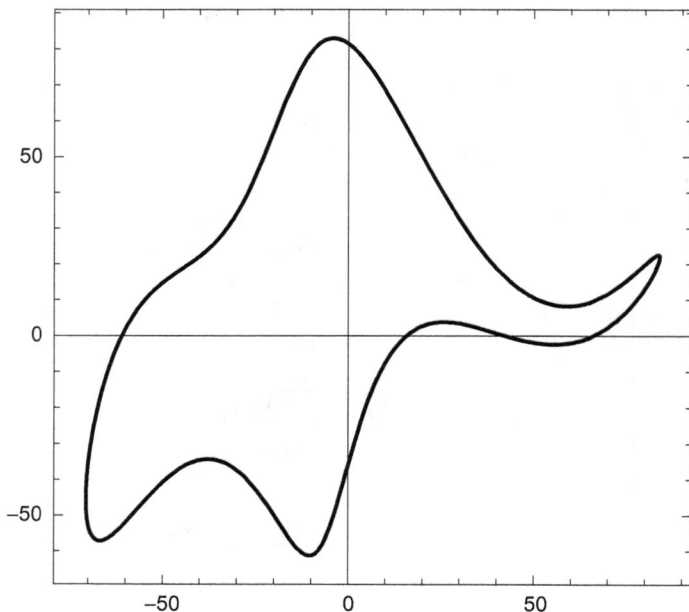

Fig. 2.2. This elephant fitted by four parameters was discovered by Mayer, Khairy, and Howard in 2010 [33].

under uncertainty, as considered in this chapter. In general, this topic is considered a subset of microeconomic theory. See Mankiw's book on the subject for details [37]. For a generalization of the results in this chapter, see Myerson's text on game theory, which provides an economic perspective [38].

$$- \spadesuit \clubsuit \heartsuit \diamondsuit -$$

2.5 Exercises

2.1 Follow the steps in Example 2.15 to construct a simple lottery describing the case when the loser decides to flip the second coin. Decide whether it is in either the winner's or the loser's interest to flip the second coin (or determine whether it is not possible to do so without more information).

2.2 Use YouTube to find a clip of the *Price is Right* game *Temptation*. Assume that you have *no* knowledge on the price of the big prize (usually a car). Model the game as a set of lotteries and compute the probabilities. [Hint: The zero knowledge assumption is critical for this to be possible.]

2.3 Explain what happens to the computation in Example 2.23 if you replace the "donkey prize" with something more unpleasant, e.g., being imprisoned for 10 years. How does this affect your acceptance of the continuity assumption?

2.4 Prove Lemma 2.25. [Hint: We know $L_1 \succeq L_2$ and $L_2 \succeq L_1$ if and only if $L_1 \sim L_2$. We also know $L_1 \succeq L_2$ if and only if $u(L_1) \geq u(L_2)$.]

2.5 Prove Theorem 2.27.

Chapter 3

Game Trees and Extensive Form

Chapter Goals: The goal of this chapter is to introduce games in extensive form, which are sometimes called game trees. This provides a visual representation for a game, similar to the one in Fig. 1.4 or Fig. 2.1. In doing this, we will slowly build more complex games from simpler ones. We begin with deterministic games of complete information (such as checkers or chess) and proceed to games of incomplete information and games with probabilistic moves. We then define and prove the existence of an equilibrium strategy in the case of complete information, and we conclude with Zermelo's theorem.

3.1 Graphs and Trees

Remark 3.1. We have already seen two diagrams (Figs. 1.4 and 2.1) with a branching structure modeling decisions that individuals can make. We now formalize these diagrams using the language of graph theory [39].

Definition 3.2 (Graph). A directed graph (digraph) is a pair $G = (V, E)$ where V is a finite set of vertexes and $E \subseteq V \times V$ is a finite set of directed edges composed of ordered two-element subsets of V.

Remark 3.3. For the sake of simplicity, we assume that edges of the form (v, v) (called self-loops) are not permissible in the graph we draw because they have no game-theoretic meaning.

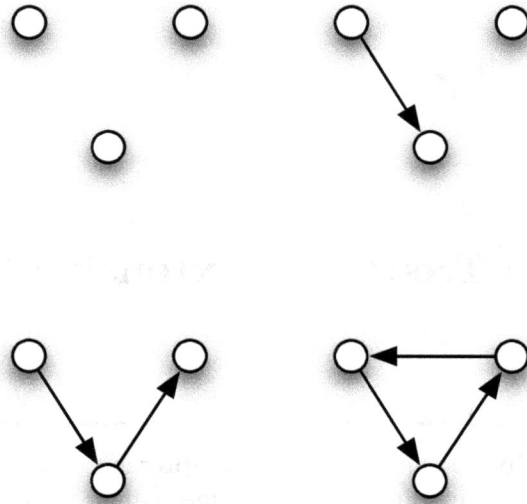

Fig. 3.1. There are $64 = 2^6$ distinct graphs on three vertices. The increased number of edges is caused by the fact that the edges are now directed.

Example 3.4. Under the previous assumption, there are $2^6 = 64$ possible digraphs on three vertices. This can be computed by considering the number of permutations of two elements chosen from a three-element set. This yields six possible ordered pairs of vertices (directed edges). For each of these edges, there are two possibilities: either the edge is in the edge set or not. Thus, the total number of digraphs on three edges is $2^6 = 64$. A few are illustrated in Fig. 3.1.

Definition 3.5 (Directed Path). Let $G = (V, E)$ be a digraph. Then, a *directed path* in G is a sequence of vertices $(v_0, v_1, \ldots, v_n)$ so that $(v_i, v_{i+1}) \in E$ for each $i = 0, \ldots, n-1$ and no vertex is repeated. We say that the path goes from vertex v_0 to vertex v_n. The number of edges in a path is called its *length*.

Example 3.6. We illustrate a short path in Fig. 3.2. The directed edges in this graph prevent many long paths.

Definition 3.7 (Directed Tree). A digraph $G = (V, E)$ that possesses a unique vertex $r \in V$ called the *root* so that (i) there is a

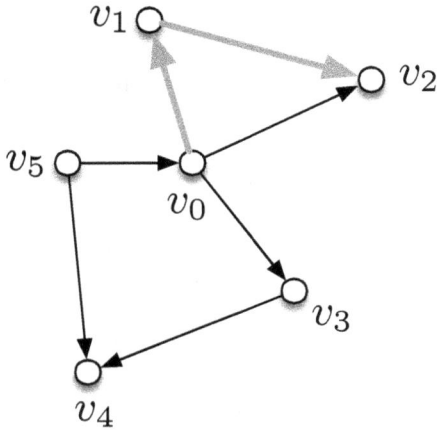

Fig. 3.2. A short path consisting of three vertices is illustrated in a directed graph.

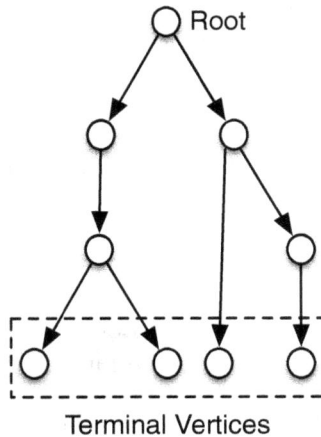

Fig. 3.3. We illustrate a directed tree. Every directed tree has a unique vertex called the *root*. The root is connected by a unique directed path to every other vertex in the directed tree.

unique path from r to every vertex $v \in V$ and (ii) there is no $v \in V$ so that $(v, r) \in E$ is called a *directed tree*.

Example 3.8. Figure 3.3 illustrates a simple directed tree. Note that there is exactly one directed path connecting the root to every other vertex in the tree.

Remark 3.9. We note that there are other ways to define trees (including directed trees). This one is most convenient for us because we will always be able to identify an obvious root vertex. See Ref. [39] for additional information on graphs and trees.

Definition 3.10 (Descendants). If $T = (V, E)$ is a directed tree and $v, u \in V$ with $(v, u) \in E$, then u is called a *child* of v, and v is called the *parent* of u. If there is a path from v to u in T, then u is called a *descendant* of v, and v is called an *ancestor* of u.

Definition 3.11 (Out-Edges). If $T = (V, E)$ is a directed tree and $v \in V$, then we denote the *out-edges* of vertex v by $E_o(v)$. These are edges that connect v to its children. Thus,

$$E_o(v) = \{(v, u) \in V : (v, u) \in E\}.$$

Definition 3.12 (Terminal Vertex). If $T = (V, E)$ is a directed tree and $v \in V$ so that v has no descendants, then v is called a *terminal* or *terminating* vertex. All vertices that are not terminal are *non-terminal* or *intermediate* vertices.

Remark 3.13. In other contexts, terminal vertices are called leaves of the tree, and a terminal vertex is called a leaf [39]. Because digraphs (and more generally graphs) are used in many disciplines, a wide range of terminology is in use, depending on the context.

Definition 3.14 (Tree Height). Let $T = (V, E)$ be a tree. The *height* of the tree is the length of the longest path in T.

Example 3.15. The height of the tree shown in Fig. 3.3 is 3. There are three paths of length 3 in the tree that start at the root of the tree and lead to three of the four terminal vertices.

Lemma 3.16. *Let $T = (V, E)$ be a directed tree. If v is a vertex of v and u is a descendant of v, then there is no path from u to v.*

Proof. Let r be the root of the tree. If $v = r$, then the theorem is proved. Suppose not. Let $(w_0, w_1, \ldots, w_n)$ be a path from u to v, with $w_0 = u$ and $w_n = v$. Let $(x_0, x_1, \ldots, x_m)$ be the path from the root of the tree to the node v (thus $x_0 = r$ and $x_m = v$). Let $(y_0, y_1, \ldots, y_k)$

be the path leading from r to u (thus $y_0 = r$ and $y_k = u$). Then, we can construct a new path,

$$\langle r = y_0, y_1, \ldots, y_k = u = w_0, w_1, \ldots, w_n = v \rangle,$$

from r (the root) to the vertex v. Thus, there are two paths leading from the root to vertex v, contradicting our assertion that T was a tree. □

Theorem 3.17. *Let $T = (V, E)$ be a tree. Suppose $u \in V$ is a vertex, and let*

$$V(u) = \{v \in V : v = u \text{ or } v \text{ is a descendant of } u\}.$$

Let $E(u)$ be the set of all edges defined in paths connecting u to a vertex in $V(u)$. Then, the graph $T_u = (V(u), E(u))$ is a tree with root u and is called the subtree of T descended from u.

Example 3.18. A subtree of the tree in Example 3.8 is shown in Fig. 3.4. Subtrees can be useful in analyzing decisions in games.

Proof of Theorem 3.17. If u is the root of T, then the statement is clear. There is a unique path from u (the root) to every vertex in T, by definition. Thus, T_u is the whole tree.

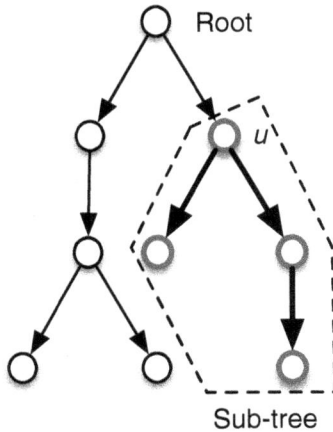

Fig. 3.4. We illustrate a subtree. This tree is the collection of all nodes that are descended from a vertex u.

Suppose that u is not the root of T. The set $V(u)$ consists of all descendants of u and u itself. Thus, between u and each $v \in V(u)$, there is a path $p = \langle v_0, v_1, \ldots, v_n \rangle$ where $v_0 = u$ and $v_n = v$. To see that this path must be unique, suppose that it is not. Then, there is at least one other distinct path $(w_0, w_1, \ldots, w_m)$ with $w_0 = u$ and $w_m = v$. But if that's so, we know there is a unique path $(x_0, \ldots, x_k)$ with x_0 being the root r of T and $x_k = u$. It follows that there are two paths,

$$(r = x_0, \ldots, x_k = v_0 = u, v_1, \ldots, v_n = v) \quad \text{and}$$

$$(r = x_0, \ldots, x_k = w_0 = u, w_1, \ldots, w_m = v),$$

between the root x_0 and the vertex v. This is a contradiction of our assumption that T was a directed tree.

To see that there is no path leading from any element in $V(u)$ back to u, we apply Lemma 3.16. Since, by definition, every edge found in the paths connecting u with its descendants is in $E(u)$, it follows that T_u is a directed tree and u is the root since there is a unique path from u to each element of $V(u)$ and there is no path leading from any element of $V(u)$ back to u. This completes the proof. $\square$

3.2   Game Trees with Complete Information and No Chance

Remark 3.19. We now define a special type of *game tree* with *perfect information* and *no chance moves*. For this, we consider a player set $\mathbf{P} = \{P_1, \ldots, P_N\}$ and a finite set of allowable moves $\mathcal{S}$ that the players can make. Formally, players and moves are just labels to be assigned to the vertices and edges of a tree. Terminal vertices in the tree correspond to end-game conditions (e.g., checkmate in chess), and each terminal vertex is assigned a payoff (score or prize). We start with the assumption that there are no "chance moves" and all players "know" precisely who is allowed to move and what moves were made. By this, we mean that when reasoning about player decisions, we assume players have complete knowledge of the game tree and any prior moves made. We introduce chance moves later.

Definition 3.20 (Player Vertex Assignment). Let $T = (V, E)$ be a directed tree, with $F \subseteq V$ the terminal vertices of T and D

the non-terminal vertices of T. (See Definition 3.12.) An *assignment of players to vertices* is an onto function $\nu : D \to \mathbf{P}$ that assigns to each non-terminal vertex $v \in D$ a player $\nu(v) \in \mathbf{P}$. Player $\nu(v)$ is said to *own* or *control* vertex v.

Remark 3.21. In the context of game trees, the non-terminal vertices of a game tree are sometimes called *decision vertices*, which is why we use D to represent them in Definition 3.20.

Definition 3.22 (Move Assignment). Let $T = (V, E)$ be a directed tree. A *move assignment function* is a mapping $\mu : E \to \mathcal{S}$ where $\mathcal{S}$ is a finite set of player moves, where we enforce the condition that if $v, u_1, u_2 \in V$ and $(v, u_1) \in E$ and $(v, u_2) \in E$, then $\mu(v, u_1) = \mu(v, u_2)$ if and only if $u_1 = u_2$.

Remark 3.23. The condition given in Definition 3.22 simply states that we must label each out-edge from a vertex v with a unique move. That is, there cannot be any question in the player's mind about what will happen when she chooses a move.

Definition 3.24 (Payoff Function). If $T = (V, E)$ is a directed tree, let $F \subseteq V$ be the terminal vertices. A payoff function is a mapping $\pi : F \to \mathbb{R}^N$ that assigns to each terminal vertex of T a numerical payoff for each player in $\mathbf{P} = \{P_1, \ldots, P_N\}$.

Remark 3.25. It is possible that the payoffs from a game may not be real-valued but instead be tangible assets, prizes, or penalties. We assume that the assumptions of the expected utility theorem (Theorem 2.21) are in force; therefore, a linear utility function can be defined that provides the real values required for the definition of the payoff function π.

Definition 3.26 (Game Tree with Complete Information and No Chance Moves). A *game tree with complete information and no chance moves* is a tuple (ordered list), $\mathcal{G} = (T, \mathbf{P}, \mathcal{S}, \nu, \mu, \pi)$, such that T is a directed tree, $\mathbf{P}$ is the set of players, $\mathcal{S}$ is the set of moves, ν is a player vertex assignment on intermediate vertices of T, μ is a move assignment on the edges of T, and π is a payoff function on T.

Example 3.27 (Rock-Paper-Scissors). Consider an odd version of rock-paper-scissors played between two people, in which Player 1

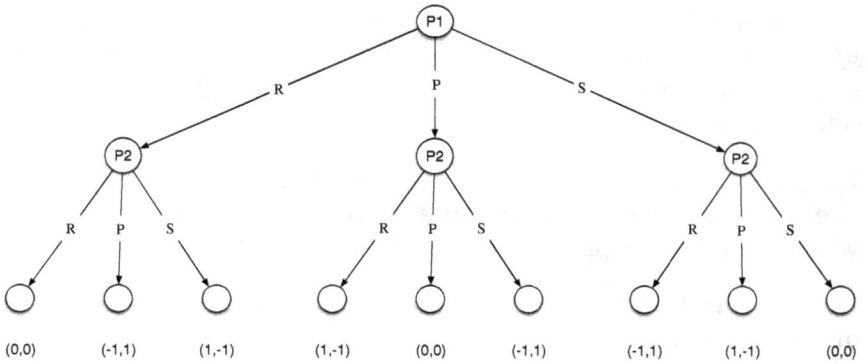

Fig. 3.5. Rock-paper-scissors with perfect information: Player 1 moves first and holds up a symbol for either rock, paper, or scissors. This is illustrated by the three edges leaving the root node, which is assigned to Player 1. Player 2 then holds up a symbol for either rock, paper, or scissors. Payoffs are assigned to Players 1 and 2 at terminal nodes. The index of the payoff vector corresponds to the players.

throws and then Player 2 throws. If we assume that the winner receives $+1$ points and the loser receives -1 points (and in ties, both players win 0 points), then the game tree for this scenario is shown in Fig. 3.5. You may think this game is not entirely *fair*, which is not mathematically defined, because it looks like Player 2 has an advantage in knowing Player 1's move before making his own move. Irrespective of this feeling, this is a valid game tree.

Definition 3.28 (Strategy-Perfect Information). Let $\mathcal{G} = (T, \mathbf{P}, \mathcal{S}, \nu, \mu, \pi)$ be a game tree with complete information and no chance, with $T = (V, E)$. Let $V_i \subset V$ be the vertices controlled by Player i. A *pure strategy* for player P_i (in a perfect information game) is a mapping $\sigma_i : V_i \to \mathcal{S}$ with the property that if $v \in V_i$ and $\sigma_i(v) = s$, then there is some $u \in V$ so that $(v, u) \in E$ and $\mu(v, u) = s$. (Thus, σ_i will only choose a move that labels an edge leaving v.)

Remark 3.29. Definition 3.28 tells us that a pure strategy for player P_i is to choose one out-edge from each vertex controlled by that player. This is the move the player will make at that point in the game.

Remark 3.30 (Rationality). As we build methods to find strategies for players, we assume that players are *rational*, that at any time they know the *entire* game tree, and that each player will attempt to maximize her payoff at the end of the game by choosing an appropriate strategy function σ_i in reference to all the strategy functions any other player might choose.

Example 3.31 (The Battle of the Bismark Sea – Part 1[1]). Games can be used to illustrate the importance of intelligence in combat. In February 1943, the battle for New Guinea had reached a critical juncture in World War II. The Allies controlled the southern half of New Guinea and the Japanese the northern half. Reports indicated that the Japanese were amassing troops to reinforce their army in New Guinea in an attempt to control the entire island. These troops had to be delivered by a naval convoy. The Japanese had a choice of sailing either north of New Britain, where rain and poor visibility were expected, or sailing south of New Britain, where the weather was expected to be good. Either route required the same amount of sailing time. See Fig. 3.6.

General Kenney, the Allied forces commander in the Southwest Pacific, had been ordered to do as much damage to the Japanese convoy fleet as possible. He had reconnaissance aircraft to detect the Japanese fleet but had to determine whether to concentrate his search planes on the northern or southern route.

The game tree in Fig. 3.7 summarizes the choices for the Japanese (J) and American (A) commanders (players), with payoffs given as the *number of days available for the bombing of the Japanese fleet*. (Since the Japanese cannot benefit, their payoff is reported as the negative of these values.) The moves for each player are to either *sail north* or *sail south* for the Japanese and to either *search north* or *search south* for the Americans.

In this game tree, we assume perfect information. Thus, the Americans (somehow) *know* which route the Japanese will sail. Knowing this, they can make an optimal choice for each contingency. If the Japanese sail north, then the Americans search north and will be able to bomb the Japanese fleet for two days. Similarly, if the

---

[1]This example is discussed in detail by Brams in Ref. [40].

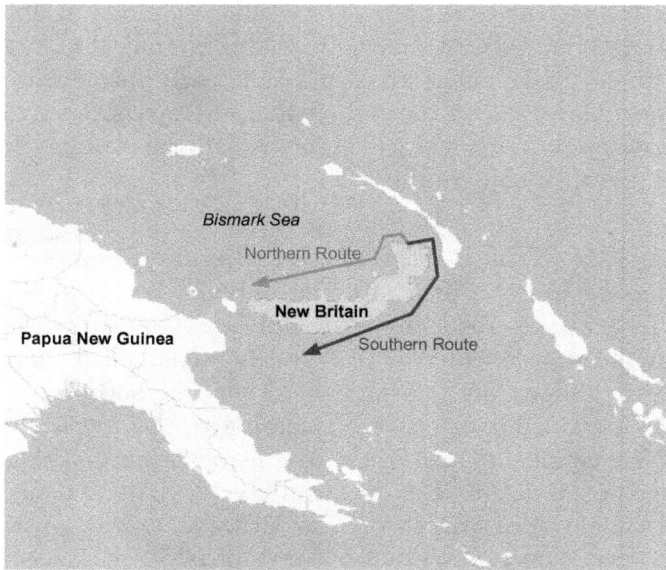

Fig. 3.6.   New Guinea is located in the South Pacific and was a major region of contention during World War II. The northern half was controlled by Japan until 1943, while the southern half was controlled by the Allies. The routes shown are approximate.

Japanese sail south, the Americans will search south and be able to bomb the Japanese fleet for three days.

The Japanese, however, also have access to this game tree and, reasoning that the Americans are payoff maximizers, will choose a path to minimize their exposure to attack. They *must* choose to go north and accept two days of bombing. If they choose to go south, then they know they will be exposed to three days of bombing. Thus, their optimal strategy is to sail north.

Naturally, the Allies did not know which route the Japanese would take. We will return to this case later.

Remark 3.32.   The complexity of a game (especially one with perfect information and no chance moves) can often be measured by how many nodes are in its game tree. Certain games, such as chess and Go, have *huge* game trees. When computers play games, they often attempt to explore a game tree in order to determine optimal moves (even when using neural networks, as in AlphaGo [41]).

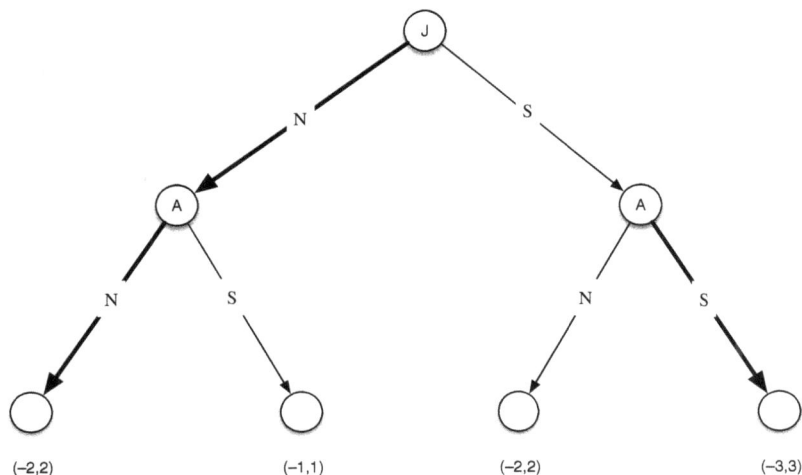

Fig. 3.7. The Japanese could choose to sail either north or south of New Britain. The Americans (Allies) could choose to concentrate their search efforts on either the northern or southern routes. Given this game tree, the Americans would always choose to search the north if they *knew* the Japanese had chosen to sail on the north side of New Britain. Alternatively, they would search the south route if they knew the Japanese had taken that route. Assuming the Americans had perfect intelligence, the Japanese would always choose to sail the northern route, as they would expose themselves to only two days of bombing as opposed to three with the southern route.

Another measure of complexity is the length of the longest path in the game tree. In Example 3.31, the length of the longest path in the game tree is two edges (moves) containing three nodes. This reflects the fact that there are only two moves in the game: first, Player 1 moves, and then, Player 2 moves.

3.3 Game Trees with Incomplete Information

Remark 3.33 (Power Set and Partitions). Recall from Remark 1.15 that, if X is a set, then 2^X is the power set of X or the set of all subsets of X.

Definition 3.34 (Partition). Let X be a set. A partition of X is a set $\mathcal{I} \subseteq 2^X$ so that for all $x \in X$, there is exactly one element $I \in \mathcal{I}$ so that $x \in I$.

Definition 3.35 (Information Sets). If $T = (V, E)$ is a tree and $D \subset V$ are the intermediate (decision) nodes of the tree, ν is a player assignment function, and μ is a move assignment, then *information sets* are a set of subsets $\mathcal{I} \subset 2^D$ satisfying the following:

(1) For all $v \in D$, there is exactly one set $I_v \in \mathcal{I}$ such that $v \in I_v$. This is the information set of the vertex v.
(2) If $v_1, v_2 \in I_v$, then $\nu(v_1) = \nu(v_2)$.
(3) If $(v_1, v) \in E$ and $\mu(v_1, v) = m$, and $v_2 \in I_{v_1}$ (i.e., v_1 and v_2 are in the same information set), then there is some $w \in V$ so that $(v_2, w) \in E$ and $\mu(v_2, w) = m$.

Thus, $\mathcal{I}$ is a partition of D.

Remark 3.36. Definition 3.35 says that every vertex in a game tree is assigned to one information set. It also says that if two vertices are in the same information set, then they must both be controlled by the same player. Finally, the definition says that two vertices can be in the same information set only if the moves from these vertices are indistinguishable.

Remark 3.37. An information set is used to capture the notion that a player doesn't know what vertex of the game tree she is at, i.e., she cannot distinguish between two nodes in the game tree. All that is known is that the same moves are available at all vertices in a given information set.

In a case like this, it is possible that the player doesn't know which vertex in the game tree will come next as a result of choosing a move, but she can certainly limit the number of possible vertices.

Remark 3.38. We can also think of the information set as a mapping $\xi : V \to \mathcal{I}$, where $\mathcal{I}$ is a finite set of information labels and the labels satisfy requirements like those in Definition 3.35.

Definition 3.39 (Game Tree with Incomplete Information and No Chance Moves). A *game tree with incomplete information and no chance* is a tuple, $\mathcal{G} = (T, \mathbf{P}, \mathcal{S}, \nu, \mu, \pi, \mathcal{I})$, such that T is a directed tree, ν is a player vertex assignment on intermediate vertices of T, μ is a move assignment on the edges of T, and π is a payoff function on T, and $\mathcal{I}$ are information sets.

Definition 3.40 (Strategy). Let $\mathcal{G} = (T, \mathbf{P}, \mathcal{S}, \nu, \mu, \pi, \mathcal{I})$ be a game tree with incomplete information and no chance moves, with $T = (V, E)$. Let $\mathcal{I}_i$ be the information sets controlled by Player i. A *pure strategy* for Player P_i is a mapping $\sigma_i : \mathcal{I}_i \to \mathcal{S}$ with the property that if $I \in \mathcal{I}_i$ and $\sigma_i(I) = s$, then for every $v \in I$, there is some edge $(v, w) \in E$ so that $\mu(v, w) = s$. The set of all strategies for player i is denoted by Σ_i.

Proposition 3.41. *If $\mathcal{G} = (T, \mathbf{P}, \mathcal{S}, \nu, \mu, \pi, \mathcal{I})$ and $\mathcal{I}$ consists of only singleton sets, then $\mathcal{G}$ is equivalent to a game with complete information.*

Proof. The information sets are used only for defining strategies. Since each $I \in \mathcal{I}$ is a singleton, we know that for each $I \in \mathcal{I}$, we have $I = \{v\}$, where $v \in D$. Here, D is the set of decision nodes in V, with $T = (V, E)$. Thus, any strategy $\sigma_i : \mathcal{I}_i \to E$ can easily be converted into $\sigma_i : V_i \to E$ by stating that $\sigma_i(v) = \sigma_i(\{v\})$ for all $v \in V_i$. This completes the proof. $\qquad\square$

Remark 3.42. Note that Definition 3.40 fully generalizes Definition 3.28 since, in a perfect information game, each decision vertex is considered to be in its own information set.

Example 3.43 (The Battle of the Bismark Sea – Part 2). Recall Example 3.31. Obviously, General Kenney did *not* know *a priori* which route the Japanese would take. This can be modeled using information sets. In this game, the two nodes that are owned by the Allies in the game tree are in the same information set. General Kenney doesn't know whether the Japanese will sail north or south. He could (in theory) have reasoned that they should sail north, but he doesn't know. The information set for the Japanese is likewise shown in Fig. 3.8 as dashed rectangles.

From the perspective of the Japanese, since the routes will take the same amount of time, the northern route is more favorable. To see this, consider Table 3.1.

If the Japanese sail north, then the worst they will suffer is two days of bombing, while the best they will suffer is one day of bombing. If the Japanese sail south, the worst they will suffer is three days of bombing, while the best they will suffer is two days of bombing.

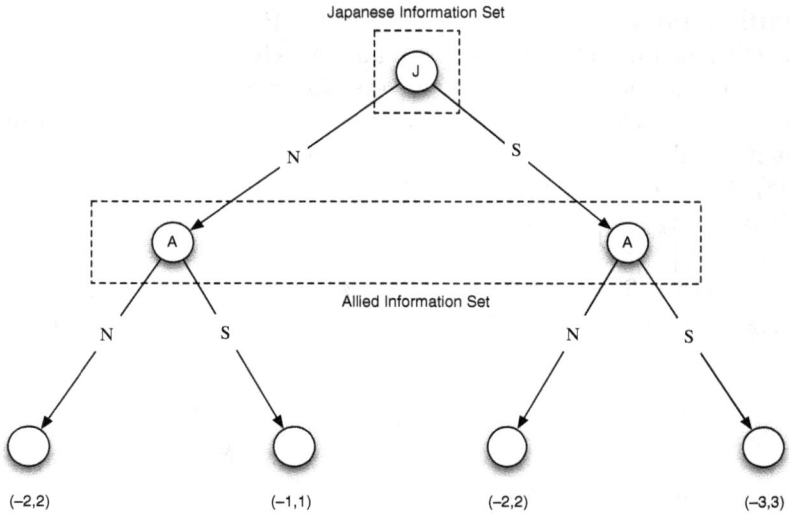

Fig. 3.8. The game tree for the Battle of the Bismark Sea with incomplete information. Obviously, Kenney could not have known *a priori* which path the Japanese would choose to sail. He could have reasoned (as they might have) that their best plan was to sail north, but he wouldn't really *know*. We can capture this fact by showing that when Kenney chooses his move, he cannot distinguish between the two intermediate nodes that belong to the Allies.

Table 3.1. Various strategies and payoffs in the Battle of the Bismark Sea. The northern route is favored by the Japanese, who will always do no worse in taking it than they do the southern route.

	Sail North		**Sail South**
Search North	Bombed for 2 days	$\leq$	Bombed for 2 days
Search South	Bombed for 1 day	$\leq$	Bombed for 3 days

Thus, the northern route should be preferable, as the cost of taking it is never worse than taking the southern route. We say that the northern route strategy *dominates* the southern route strategy. (We discuss this more formally later.) If General Kenney could reason this out, then he might choose to commit his reconnaissance forces to searching the north, even without being able to determine whether the Japanese sailed north or south.

3.4 Games of Chance

Remark 3.44. In games of chance, there is always a point in the game where a chance move is made. In card games, the initial deal is one of these points. To accommodate chance moves, we assume the existence of a *Player 0*, who is sometimes called *Nature*. When dealing with games of chance, we assume that the player vertex assignment function assigns some vertices the label P_0.

Definition 3.45 (Moves of Player 0). Let $T = (V, E)$ and ν be a player vertex assignment function. For all $v \in D$ such that $\nu(v) = P_0$, there is a probability assignment function $p_v : E_o(v) \to [0, 1]$ satisfying

$$\sum_{e \in E_o(v)} p_v(e) = 1. \tag{3.1}$$

Remark 3.46. The probability function(s) p_v in Definition 3.45 essentially defines a roll of the dice. When game play reaches a vertex owned by P_0, Nature (or Player 0 or Chance) probabilistically advances the game by moving along a randomly chosen edge. The fact that Eq. (3.1) holds simply asserts that the chance moves of Nature form a probability space at that point, whose outcomes are all the possible chance moves.

Definition 3.47 (Game Tree). Let $T = (V, E)$ be a directed tree, $F \subseteq V$ be the terminal vertices, and $D = V \backslash F$ be the intermediate (or decision) vertices. Let $\mathbf{P} = \{P_0, P_1, \ldots, P_N\}$ be a set of players, including the chance player P_0. Let $\mathcal{S}$ be a set of moves for the players. Let $\nu : D \to \mathbf{P}$ be a player vertex assignment function and $\mu : E \to \mathcal{S}$ be a move assignment function. Let $\mathcal{P}$ denote the set of Player 0 move functions. That is,

$$\mathcal{P} = \{p_v : \nu(v) = P_0\}.$$

Let $\pi : F \to \mathbb{R}^N$ be a payoff function. Let $\mathcal{I} \subseteq 2^D$ be the set of information sets.

A *game tree* is a tuple: $\mathcal{G} = (T, \mathbf{P}, \mathcal{S}, \nu, \mu, \pi, \mathcal{I}, \mathcal{P})$. In this form, the game defined by the game tree $\mathcal{G}$ is said to be in *extensive form*.

Remark 3.48. A strategy for Player i in a game tree like the one in Definition 3.47 is the same as that in Definition 3.40.

Example 3.49 (Coin Flip Poker). At the beginning of this game, each player antes up \$1 into a common pot. Player 1 flips a coin and knows its outcome. Player 2 does not. Player 1 has the option of passing or raising:

(1) If Player 1 passes, she shows the coin to Player 2; if the coin is heads, Player 1 wins the pot, whereas if the coin is tails, Player 1 loses the pot.
(2) If Player 1 raises, then she adds another dollar to the pot, and Player 2 must decide whether to call or fold:
   (a) If Player 2 folds, then the game ends and Player 1 takes the money, irrespective of the coin.
   (b) If Player 2 calls, then he adds \$1 to the pot. Player 1 shows the coin; if the coin shows heads, then she wins the pot (\$2) and Player 2 loses the pot, whereas if the coin shows tails, then she loses the pot and Player 2 wins the pot (\$2).

The game tree for this game is shown in Fig. 3.9. The root node of the game tree is controlled by Nature (Player 0). This corresponds to the initial coin flip of Player 1, which is random and will result in heads 50% of the time and tails 50% of the time.

Note that the nodes controlled by P_2 are in the same information set. This is because it is *impossible* for Player 2 to know whether Player 1 has flipped heads or tails.

The payoffs shown on the terminal nodes are determined by how much each player will win or lose.

3.5   Payoff Functions and Equilibria

Remark 3.50 (Some Notation). To study equilibria, we have to work with subtrees. Unfortunately, the notation can get cumbersome, but it will drive all our proofs.

Let $\mathcal{G} = (T, \mathbf{P}, \mathcal{S}, \nu, \mu, \pi, \mathcal{I}, \mathcal{P})$ be a game tree, and let $u \in D$, where D is the set of non-terminal vertices of T. Recall that T_u is the subtree of T rooted at u. Then, $V(T_u)$ denotes the vertex set of T_u. Suppose $\nu : D \to \mathbf{P}$ is a player vertex assignment function, then

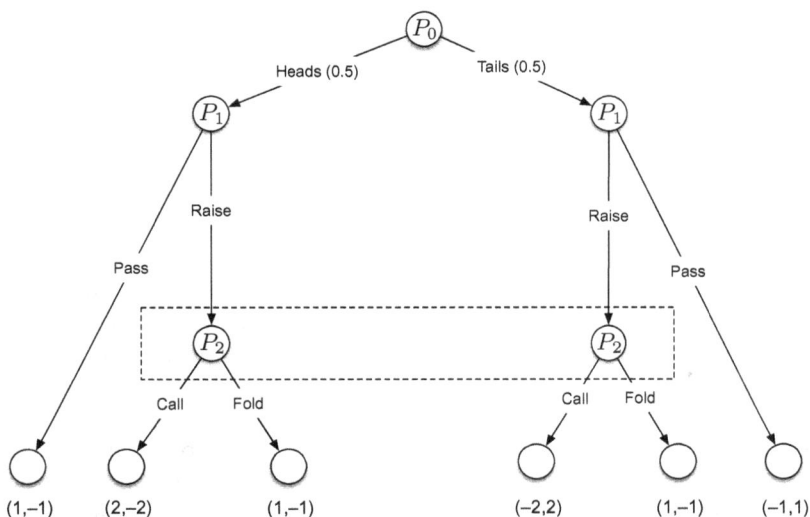

Fig. 3.9. The root node of the game tree is controlled by Nature. Player 1's coin flip happens at this point. Player 1 can then decide whether to end the game by passing (and thus receiving a payoff or not) or continue the game by raising. At this point, Player 2 can then decide whether to call or fold, thus potentially receiving a payoff.

$\nu|_{T_u} : V(T_u) \cap D \to \mathbf{P}$ is the function ν *restricted* to the vertices in the subtree. Other functions restricted to T_u are denoted similarly.

Theorem 3.51. *Let* $\mathcal{G} = (T, \mathbf{P}, \mathcal{S}, \nu, \mu, \pi, \mathcal{I}, \mathcal{P})$ *be a game tree, and let* $u \in D$, *where* D *is the set of non-terminal vertices of* T. *Then, the following is a game tree:*

$$\mathcal{G}' = (T_u, \mathbf{P}, \mathcal{S}, \nu|_{T_u}, \mu|_{T_u}, \pi|_{T_u}, \mathcal{I}|_{T_u}, \mathcal{P}|_{T_u}),$$

where $\mathcal{I}|_{T_u}$ *is the set of information sets restricted to the vertices of* T_u. *That is,*

$$\mathcal{I}|_{T_u} = \{I \cap V(T_u) : I \in \mathcal{I}\}.$$

Finally, $\mathcal{P}|_{T_u}$ *is the set of probability assignment functions in* $\mathcal{P}$ *restricted only to the edges in* T_u.

Proof. By Theorem 3.17, we know that T_u is a subtree of T. The functions $\nu|_{T_u}$, $\mu|_{T_u}$, and $\pi|_{T_u}$ are simply restrictions of the respective functions to the subtree.

Let v be a descendant of u controlled by chance. Since all descendants of u are included in T_u, it follows that all descendants of v are contained in T_u. Thus,

$$\sum_{e \in E_o(v)} p_v|_{T_u}(e) = 1,$$

as required. Thus, $\mathcal{P}|_{T_u}$ is an appropriate set of probability functions.

Finally, since $\mathcal{I}$ is a partition of T_u, we may compute $\mathcal{I}|_{T_u}$ by simply removing the vertices in the subsets of $\mathcal{I}$ that are not in T_u. This set, $\mathcal{I}_{T_u}$, is a partition of T_u and necessarily satisfies the requirements set forth in Definition 3.35 because all the descendants of u are elements of $V(T_u)$. $\qquad\square$

Example 3.52. If we consider the game in Example 3.49, but suppose that Player 1 is known to have flipped heads, then the new game tree is derived by considering only the subtree in which Player 1 sees heads. This is shown in Fig. 3.10. It is worth noting that when we restrict our attention to this subtree, a game that was originally an incomplete-information game becomes a complete-information game. That is, each vertex is now the sole member of its information set. Additionally, we have removed chance from the game.

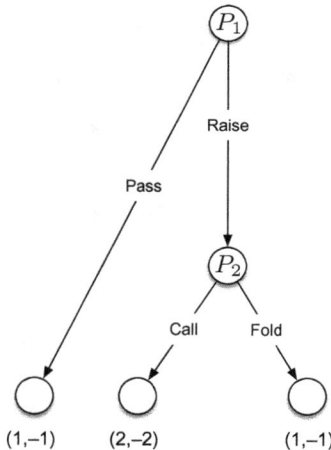

Fig. 3.10. We are told that Player 1 flips heads. The resulting game tree is substantially simpler. Because the information set on Player 2's controlled nodes indicated a lack of knowledge of Player 1's card, we can see that this sub-game is now a complete-information game.

Theorem 3.53. *Let* $\mathcal{G} = (T, \mathbf{P}, \mathcal{S}, \nu, \mu, \pi, \mathcal{I})$ *be a game with no chance. Let* $\sigma_1, \ldots, \sigma_N$ *be a set of strategies for Players* $1 - N$. *Then, these strategies determine a unique path through the game tree.*

Proof. To see this, suppose we begin at the root node r. If this node is controlled by Player i, then node r exists in the information set $I_r \in \mathcal{I}_i$. Then, $\sigma_i(I_r) = s \in \mathcal{S}$, and there is some edge $(r, u) \in E$ so that $\mu(r, u) = s$. We have a two-vertex path (r, u).

Consider the game tree $\mathcal{G}'$ constructed from subtree T_u as in Theorem 3.51. This game tree has root u. We can apply the same argument to construct a two-vertex path (u, u'), which, when joined with the initial path, forms the three-node path (r, u, u'). Repeating this argument inductively will yield a path through the game tree that is determined by the strategy functions of the players. Since the number of vertices in the tree is finite, this process will end, producing the desired path. The uniqueness of the path is ensured by the fact that the strategies are functions, and thus, at any information set, exactly one move will be chosen by the player in control. □

Example 3.54. In the Battle of the Bismark Sea, the strategy we defined in Example 3.31 clearly defines a unique path through the tree. Since each player determines *a priori* the unique edge he/she will select when confronted with a specific information set, a path through the tree can be determined from these selections. This is illustrated in Fig. 3.11.

Theorem 3.55. *Let* $\mathcal{G} = (T, \mathbf{P}, \mathcal{S}, \nu, \mu, \pi, \mathcal{I}, \mathcal{P})$. *Let* $\sigma_1, \ldots, \sigma_N$ *be a collection of strategies for Players* $1 - N$. *Then, these strategies determine a discrete probability space* $(\Omega, \mathcal{F}, P)$, *where* Ω *is a set of paths leading from the root of the tree to a subset of the terminal nodes, and if* $\omega \in \Omega$, *then* $P(\omega)$ *is the product of the probabilities of the chance moves defined by the path* ω.

Proof. We proceed inductively from the height of the tree T. Suppose the tree T has a height of 1. Then, there is only one decision vertex (the root). If that decision vertex is controlled by a player other than chance, then applying Theorem 3.53, we know that the strategies $\sigma_1, \ldots, \sigma_N$ define a unique path through the tree. The only paths in a tree of height 1 have the form $\langle r, u \rangle$, where r is the root of T and u is a terminal vertex. Thus, Ω is a singleton consisting of

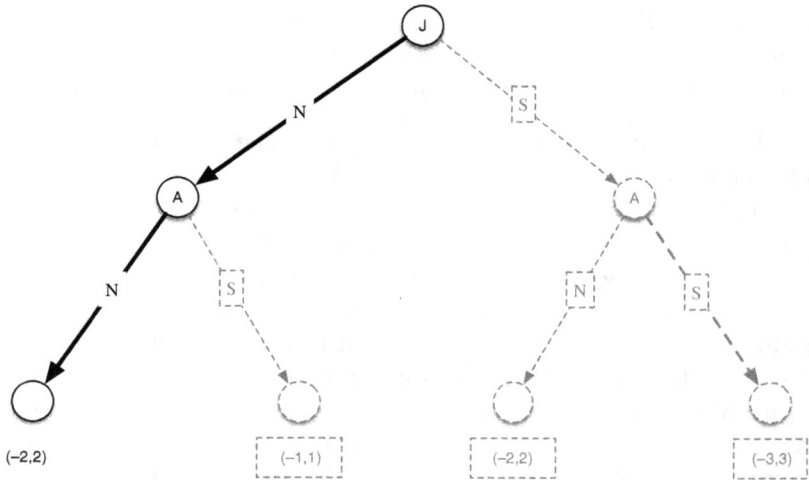

Fig. 3.11. A unique path through the game tree of the Battle of the Bismark Sea, determined by the strategies of the two players.

only the path $\langle r, u \rangle$ determined by the strategies, and it is assigned a probability of 1.

If chance controls the root vertex, then we can define

$$\Omega = \{\langle r, u \rangle : u \in F\},$$

where F is the set of terminal nodes in V. The probability assigned to a path (r, u) is simply the probability that chance (Player P_0) selects the edge $(r, u) \in E$. The fact that

$$\sum_{u \in F} p_r(r, u) = 1$$

ensures that we can define the probability space $(\Omega, \mathcal{F}, P)$. Thus, we have shown that the theorem is true for game trees of height 1.

Suppose the statement is true for game trees with heights up to $k \geq 1$. We show that the theorem is true for game trees of height $k+1$. Let r be the root of the tree T, and consider the set of children of $U = \{u \in V : (r, u) \in E\}$. For each $u \in U$, we can define a game tree of height k with tree T_u using Theorem 3.51. The fact that this tree has a height of k implies that we can define a probability space $(\Omega_u, \mathcal{F}_u, P_u)$, with Ω_u composed of paths from u to the terminal vertices of T_u.

Suppose that vertex r is controlled by Player P_j ($j \neq 0$). Then, the strategy σ_j determines a unique move that will be made by Player j at vertex r. Suppose that move m is specified by σ_j at vertex r so that $\mu(r, u) = m$ for edge $(r, u) \in E$ with $u \in U$. That is, edge (r, u) is labeled m. We define the new event set Ω of paths in the tree T from root r to a terminal vertex. The probability function on paths is defined as

$$P[(r, u_1, \ldots, u_k)] = \begin{cases} P_u[(u_1, \ldots, u_k)] & (u_1, \ldots, u_k) \in \Omega_u \text{ and } u_1 = u, \\ 0 & \text{otherwise.} \end{cases}$$

The fact that P_u is a properly defined probability function over Ω_u implies that P is a properly defined probability function over Ω, and thus $(\Omega, \mathcal{F}, P)$ is a probability space over the paths in T.

Now, suppose that chance (Player P_0) controls r in the game tree. Again, Ω is the set of paths leading from r to a terminal vertex of T. The probability function on paths can then be defined as

$$P[(r, v_1, \ldots, v_k)] = p_r(r, v_1)P_{v_1}[(v_1, \ldots, v_k)],$$

where $v_1 \in U$ and $\langle v_1, \ldots, v_k \rangle \in \Omega_{v_1}$, the set of paths leading from v_1 to a terminal vertex in the tree T_{v_1}, and $p(r, v_1)$ is the probability chance assigned to the edge $(r, v_1) \in E$.

To see that this is a properly defined probability function, suppose that $\omega \in \Omega_u$. That is, ω is a path in the tree T_u leading from u to a terminal vertex of T_u. Then, a path in Ω is constructed by joining the path that leads from vertex r to vertex u and then following a path $\omega \in \Omega_u$. Let (r, ω) denote such a path. Then, we know that

$$\sum_{u \in U} \sum_{\omega \in \Omega_u} P[(r, \omega)] = \sum_{u \in U} \sum_{\omega \in \Omega_u} p(r, u)P_u(\omega)$$

$$= \sum_{u \in U} p(r, u) \left(\sum_{u \in \Omega_u} P_u(\omega) \right) = \sum_{u \in U} p(r, u) = 1.$$

$$(3.2)$$

This is because $\sum_{\omega \in \Omega_u} P_u(\omega) = 1$. Since clearly $P[(r, \omega)] \in [0, 1]$ and the paths through the game tree are independent, it follows that $(\Omega, \mathcal{F}, P)$ is a properly defined probability space. The theorem follows by induction. □

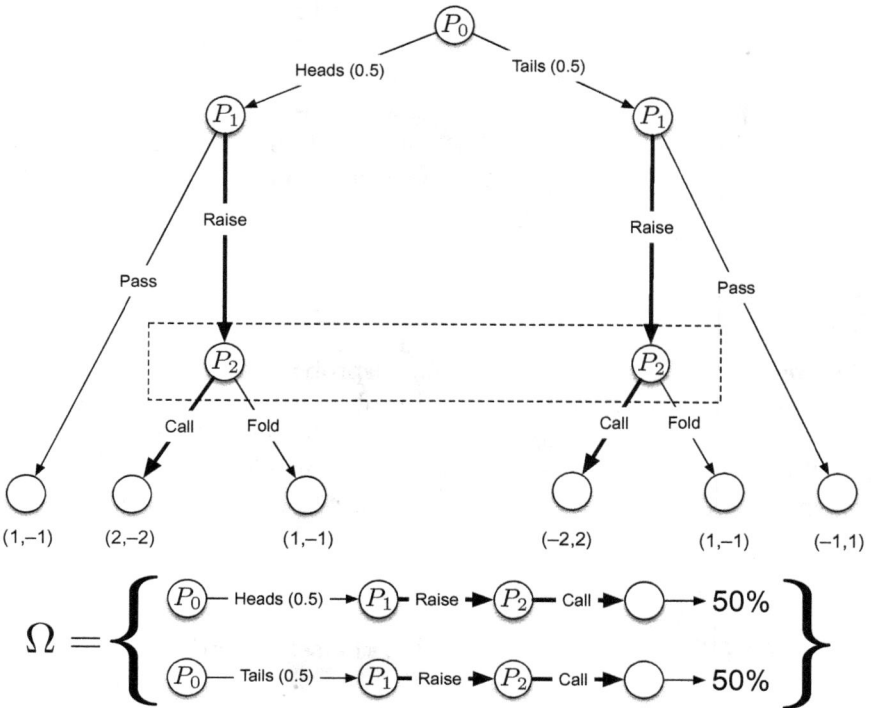

Fig. 3.12. The probability space constructed from fixed player strategies in a game of chance. The strategy space is constructed from the unique choices determined by the strategy of the players and the independent random events that are determined by the chance moves.

Example 3.56. Consider the coin-flip game from Example 3.49. Suppose we fix strategies in which Player 1 always raises and Player 2 always calls. Then, the resulting probability distribution defined as in Theorem 3.55 contains two paths: one when the coin is heads and the other when the coin is tails. This is shown in Fig. 3.12. The sample space consists of the possible paths through the game tree. As in Theorem 3.53, the paths through the game tree are completely specified because the only time probabilistic moves occur is when chance causes the game to progress.

Example 3.57. Suppose we play a game in which Players 1 and 2 ante $1 each. One card each is dealt to Player 1 and Player 2. Player 1 can choose to raise (and add $1 to the pot) or fold (and lose the pot). Player 2 can then choose to call (adding $1) or fold (and lose the pot). Player 1 wins if both cards are black. Player 2 wins if both cards are red. The pot is split if the cards have opposite colors. Suppose that Player 1 always chooses to raise and Player 2 always chooses to call. Then, the game tree and strategies are shown in Fig. 3.13. The sample space in this case consists of four distinct paths, each with a probability of $\frac{1}{4}$, assuming that the cards are dealt with equal probability. In this example, constructing the probabilities of the various events requires multiplying the probabilities of the chance moves in each path. Note that the information sets define the information that the players have. In this case, Player 1 knows the color of her card, and Player 2 knows the color of his card, but neither player knows the color of the other player's card until the game ends.

Definition 3.58 (Strategy Space). Let Σ_i be the set of all strategies for Player i in a game tree $\mathcal{G}$. Then, the entire *strategy space* is $\Sigma = \Sigma_1 \times \Sigma_2 \times \cdots \times \Sigma_n$.

Definition 3.59 (Strategy Payoff Function). Let $\mathcal{G}$ be a game tree with *no chance moves*. The *strategy payoff function* is a mapping $\pi : \Sigma \to \mathbb{R}^N$. If $\sigma_1, \ldots, \sigma_N$ are strategies for Players $1 - N$, then $\pi(\sigma_1, \ldots, \sigma_N)$ is the vector of payoffs assigned to the terminal node of the path determined by the strategies $\sigma_1, \ldots, \sigma_N$ in the game tree $\mathcal{G}$. For each $i = 1, \ldots, N$, $\pi_i(\sigma_1, \ldots, \sigma_N)$ is the payoff to Player i in $\pi_i(\sigma_1, \ldots, \sigma_N)$.

Example 3.60. Consider the Battle of the Bismark Sea game from Example 3.43. Then, there are four distinct strategies in Σ with the following payoffs:

$$\pi \,(\text{Sail North, Search North}) = (-2, 2),$$

$$\pi \,(\text{Sail South, Search North}) = (-2, 2),$$

$$\pi \,(\text{Sail North, Search South}) = (-1, 1),$$

$$\pi \,(\text{Sail South, Search South}) = (-3, 3).$$

Definition 3.61 (Expected Strategy Payoff Function). Let $\mathcal{G}$ be a game tree *with chance moves*. The *expected strategy payoff function* is a mapping $\pi : \Sigma \to \mathbb{R}^N$ defined as follows: If $\sigma_1, \ldots, \sigma_N$

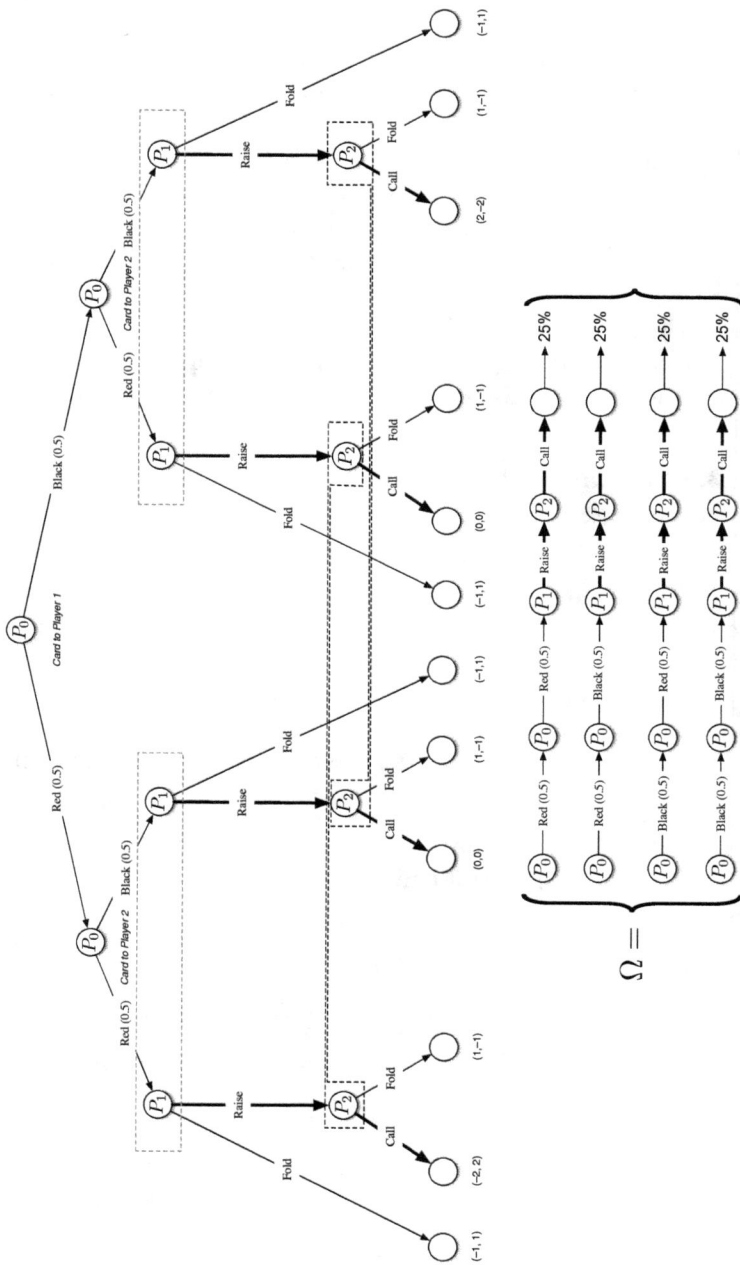

Fig. 3.13.   The probability space constructed from fixed player strategies in a game of chance. The strategy space is constructed from the unique choices determined by the strategy of the players and the independent random events that are determined by the chance moves. In this example, the probabilities of the various paths are constructed by multiplying the probabilities of the chance moves in each path.

are strategies for Players $1 - N$, then let $(\Omega, \mathcal{F}, P)$ be the probability space over the paths constructed by these strategies, as given in Theorem 3.55. Let Π_i be a random variable that maps $\omega \in \Omega$ to the payoff for Player i at the terminal node in the path ω. Let

$$\pi_i(\sigma_1, \ldots, \sigma_N) = \mathbb{E}(\Pi_i).$$

Then,

$$\pi(\sigma_1, \ldots, \sigma_N) = \langle \pi_1(\sigma_1, \ldots, \sigma_N), \ldots, \pi_N(\sigma_1, \ldots, \sigma_N) \rangle.$$

As before, $\pi_i(\sigma_1, \ldots, \sigma_N)$ is the expected payoff to Player i in $\pi(\sigma_1, \ldots, \sigma_N)$.

Example 3.62. Consider the coin-flipping game from Example 3.49. There are four distinct strategies in Σ:

$$\begin{cases} \text{(Pass, Call)}, \\ \text{(Pass, Fold)}, \\ \text{(Raise, Call)}, \\ \text{(Raise, Fold)}. \end{cases}$$

Note that the strategies specify moves at every vertex, even if they cannot be reached because of decisions by certain players.

Focus on the strategy (Pass, Call). Then, the resulting paths in the graph defined by these strategies are shown in Fig. 3.14. There are two paths, and we note that the decision made by Player 2 makes no difference in this case because Player 1 passes. Each path has a probability of $\frac{1}{2}$. Our random variable Π_1 maps the top path in

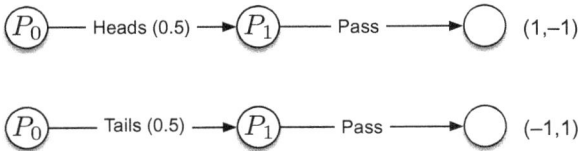

Fig. 3.14. Game tree paths derived from the game in Example 3.49 are the result of the strategy (Pass, Fold). The probability of each of these paths is $\frac{1}{2}$.

Fig. 3.14 to a \$1 payoff for Player 1 and the bottom path in Fig. 3.14 to a payoff of $-\$1$ for Player 1. Thus, we can compute

$$\pi_1 \,(\text{Pass, Call}) = \frac{1}{2}(1) + \frac{1}{2}(-1) = 0.$$

Likewise,

$$\pi_2 \,(\text{Pass, Call}) = \frac{1}{2}(-1) + \frac{1}{2}(1) = 0.$$

Thus, we compute

$$\pi \,(\text{Pass, Call}) = (0,0).$$

Using this approach, we can compute the expected payoff function to be

$$\pi \,(\text{Pass, Call}) = (0,0),$$
$$\pi \,(\text{Pass, Fold}) = (0,0),$$
$$\pi \,(\text{Raise, Call}) = (0,0),$$
$$\pi \,(\text{Raise, Fold}) = (1,-1).$$

Definition 3.63 (Equilibrium). A strategy $(\sigma_1^*, \dots, \sigma_N^*) \in \Sigma$ is an equilibrium if, for all i,

$$\pi_i(\sigma_1^*, \dots, \sigma_i^*, \dots, \sigma_N^*) \geq \pi_i(\sigma_1^*, \dots, \sigma_i, \dots, \sigma_N^*),$$

where $\sigma_i \in \Sigma_i$.

Remark 3.64. An equilibrium strategy is one in which no player can improve his/her payoff by *unilaterally* changing his/her strategy.

Example 3.65. Consider the Battle of the Bismark Sea. We can show that (Sail North, Search North) is an equilibrium strategy.

Recall that

$$\pi \left(\text{Sail North, Search North} \right) = (-2, 2).$$

Now, suppose that the Japanese deviate from this strategy and decide to sail south. Then, the new payoff is

$$\pi \left(\text{Sail South, Search North} \right) = (-2, 2).$$

Thus,

$$\pi_1 \left(\text{Sail North, Search North} \right) \geq \pi_1 \left(\text{Sail South, Search North} \right).$$

Now, suppose that the Allies deviate from the strategy and decide to search south. Then, the new payoff is

$$\pi \left(\text{Sail North, Search South} \right) = (-1, 1).$$

Thus,

$$\pi_2 \left(\text{Sail North, Search North} \right) > \pi_2 \left(\text{Sail North, Search South} \right).$$

Now, change only one player strategy at a time and evaluate whether that improves that player's payoff. In this case, neither player benefits from changing strategy.

Remark 3.66. The next proof is long and proves the existence of equilibria in games of complete information. It is safe to skip this proof, but it does introduce the idea of backward induction, which is used in dynamic programming [42], which is in turn used in more dynamic game theory.

Theorem 3.67. *Let* $\mathcal{G} = (T, \mathbf{P}, \mathcal{S}, \nu, \mu, \pi, \mathcal{I}, \mathcal{P})$ *be a game tree with complete information. Then, there is an equilibrium strategy* $(\sigma_1^*, \ldots, \sigma_N^*) \in \Sigma$.

Proof. We apply induction on the height to the game tree $T = (V, E)$. Before proceeding to the proof, recall that a game with complete information is one in which if $v \in V$ and $I_v \in \mathcal{I}$ is the information set of vertex v, then $I_v = \{v\}$. Thus, we can think of a strategy σ_i for Player P_i as a mapping from V to $\mathcal{S}$, as in Definition 3.28. We now proceed to the proof.

Suppose the height of the tree is 1. Then, the tree consists of a root node r and a collection of terminal nodes F so that if $u \in F$, then $(r, u) \in E$. If chance controls r, then there is no strategy for any of the players, and they are randomly assigned a payoff. Thus, we can think of the empty strategy as the equilibrium strategy. On the other hand, if player P_i controls r, then we let $\sigma_i(r) = m \in \mathcal{S}$ so that if $\mu(r, u) = m$ for some $u \in F$, then $\pi_i(u) \geq \pi_i(v)$ for all other $v \in F$. That is, the vertex reached by making the move m has a payoff for Player i that is greater than or equal to any other payoff Player i might receive at another vertex. All other players are assigned empty strategies (as they never make a move). Thus, it is easy to see that this is an equilibrium strategy since no player can improve their payoff by changing their strategies. Thus, we have proved that there is an equilibrium strategy in this case.

Now, suppose that the theorem is true for a game tree $\mathcal{G}$ of some height $k \geq 1$ with complete information. We show that the statement holds for game trees of height $k+1$. Let r be the root of the tree, and let $U = \{u \in V : (r, u) \in E\}$ be the set of children of r in T. If r is controlled by chance, then the first move of the game is controlled by chance. For each $u \in U$, we can construct a game tree with tree T_u by Theorem 3.51. By the induction hypothesis, we know that there is some equilibrium strategy $(\sigma_1^{u^*}, \ldots, \sigma_N^{u^*})$. Let $\pi_i^{u^*}$ be the payoff associated with using this strategy for Player P_i. Now, consider any alternative strategy $(\sigma_1^{u^*}, \ldots, \sigma_{i-1}^{u^*}, \sigma_i^u, \sigma_{i+1}^{u^*} \ldots, \sigma_N^{u^*})$. Let π_i^u be the payoff to Player P_i that results from using this new strategy in the game with game tree T_u. It must be that

$$\pi_i^{u^*} \geq \pi_i^u \quad \forall i \in \{1, \ldots, N\}, u \in U. \tag{3.3}$$

Thus, we construct a new strategy for Player P_i so that if chance causes the game to transition to vertex u in the first step, then Player P_i will use strategy $\sigma_i^{u^*}$. Eq. (3.3) ensures that Player i will never have a motivation to deviate from this strategy, as the assumption of complete information assures us that Player i will know for certain to which $u \in U$ the game has transitioned.

Alternatively, suppose that the root is controlled by Player P_j. Let U and $\pi_i^{u^*}$ be as above. Then, let $\sigma_j(r) = m \in \mathcal{S}$ so that if $\mu(r, u) = m$, then

$$\pi_j^{u^*} \geq \pi_j^{v^*} \tag{3.4}$$

for all $v \in U$. That is, Player P_j chooses a move that will yield a new game tree T_u that has the greatest terminal payoff using the equilibrium strategy $(\sigma_1^{u^*}, \ldots, \sigma_N^{u^*})$ in that game tree. We can now define a new strategy:

(1) At vertex r, $\sigma_j(r) = m$.
(2) Every move in the tree T_u is governed by $(\sigma_1^{u^*}, \ldots, \sigma_N^{u^*})$.
(3) If $v \neq r$ and $v \notin T_u$ and $\nu(v) = i$, then $\sigma_i(v)$ may be chosen at random from $\mathcal{S}$ (because this vertex will never be reached during game play).

We can show that this is an equilibrium strategy. To see this, consider any other strategy. If Player $i \neq j$ deviates, then we know that this player will receive the payoff π_i^u (as above) because Player j will force the game into the tree T_u after the first move. We know further that $\pi_i^{u^*} \geq \pi_i^u$. Thus, there is no incentive for Player P_i to deviate from the given strategy, $(\sigma_1^{u^*}, \ldots, \sigma_N^{u^*})$ in T_u.

On the other hand, suppose Player j deviates at some vertex in T_u, then we know Player j will receive the payoff $\pi_j^u \leq \pi_j^{u^*}$. Thus, once the game play takes place inside tree T_u, there is no reason to deviate from the given strategy. If Player j deviates on the first move and chooses a move m' so that $\mu(r, v) = m'$, then there are two possibilities:

(1) $\pi_j^{v^*} = \pi_j^{u^*}$,
(2) $\pi_j^{v^*} < \pi_j^{u^*}$.

In the first case, we can construct a strategy as before, in which Player P_j will still receive the same payoff as if he played the strategy in which $\sigma_j(r) = m$ (instead of $\sigma_j(r) = m'$). In the second case, the best payoff Player P_j can obtain is $\pi_j^{v^*} < \pi_j^{u^*}$, so there is certainly no reason for Player P_j to deviate by defining $\sigma_j(r) = m'$. Thus, we have shown that the strategy we constructed is an equilibrium, and it follows that there is an equilibrium strategy for this tree of height $k + 1$. The theorem follows by induction. $\square$

Remark 3.68. The equilibrium constructed in this theorem is called a sub-game perfect equilibrium because it contains an equilibrium for every possible sub-game.

Example 3.69. We can illustrate the construction from the theorem of the Battle of the Bismark Sea with complete information. In fact, you have already seen this construction once. Consider the game tree in Fig. 3.11. We construct the equilibrium solution from the bottom of the tree up. Consider the vertex controlled by the Allies, in which the Japanese sail north. In the subtree below this node, the best move for the Allies is to search north (they receive the highest payoff). This is highlighted in blue. Now, consider the vertex controlled by the Allies, where the Japanese sail south. The best move for the Allies is to search south. Now, consider the root node controlled by the Japanese. The Japanese can examine the two subtrees below this node and determine that the payoffs resulting from the equilibrium solutions in these trees are -2 (from sailing north) and -3 (from sailing south). Naturally, the Japanese will choose to make the move of sailing north, as this is the highest payoff they can achieve. The equilibrium strategy is shown in red and blue in the tree in Fig. 3.15.

Remark 3.70. We note that the equilibrium identified in the previous example also happens to be an equilibrium strategy for this game

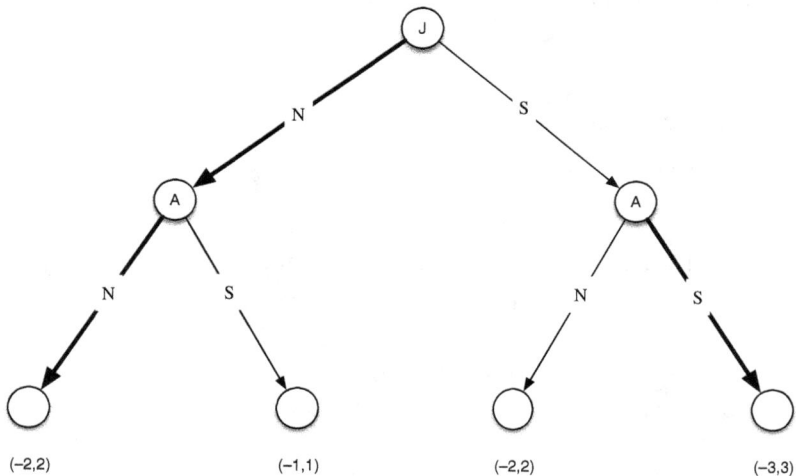

Fig. 3.15. The game tree for the Battle of the Bismark Sea. If the Japanese sail north, the best move for the Allies is to search north. If the Japanese sail south, then the best move for the Allies is to search south. The Japanese, observing the payoffs, note that given these best strategies for the Allies, their best course of action is to sail north.

if we introduce incomplete information. This is not always the case, which can be seen by investigating rock-paper-scissors with incomplete information.

Corollary 3.71 (Zermelo's Theorem). *Let $\mathcal{G} = (T, \mathbf{P}, \mathcal{S}, \nu, \mu, \pi)$ be a two-player game with complete information and no chance. Assume that the payoff is such that*

(1) *the only payoffs are +1 (win), −1 (lose);*
(2) *Player 1 wins +1 if and only if Player 2 wins −1;*
(3) *Player 2 wins +1 if and only if Player 1 wins −1.*

Finally, assume that the players take alternate turns. Then, one of the two players must have a strategy to obtain +1.

Remark 3.72. In particular, Zermelo's theorem implies that if we ensure that a chess game ends in a finite number of moves with no draws, then there is some strategy to ensure either white or black wins. We just don't know what that strategy is.

3.6   Chapter Notes

Game trees are a special case of decision trees [43], which appear frequently in artificial intelligence. In fact, game play was one of the earliest applications of artificial intelligence – in the classical sense [44]. In this case, the so-called $\alpha - \beta$ pruning [45] is used to explore a game tree, allowing a machine player to make optimal moves. Variants of this approach reached their peak with the creation of Deep Blue [46], the chess-playing algorithm that defeated Kasparov in 1996, and the solution of checkers in 2007 [47]. Games with complete information and no chance moves (such as chess and checkers) are often called combinatorial games and are covered in depth by Berlekamp, Conway, and Guy's four-volume *Winning Ways for Your Mathematical Plays* [48–51].

The challenge with game-tree analysis is the combinatorial explosion of the tree size in most games. Modern deep-learning-based methods still use a type of tree exploration (often called fictitious play) but adopt a radically different approach to computing the long-run payoff estimates compared to the methods used by Deep Blue. The approach used by Deep Mind to win at Go [52] is effectively a combination of simulation and payoff function approximation. Fu [41] offers an excellent summary of the approach from the perspective of simulation.

Game trees are extensively used in economics, where they are sometimes referred to as dynamic games – though this is not universal. Myerson [38] provides a detailed introduction to game theory from an economics perspective and refers to these games as being in extensive form, as we do here. A classic example of an economic game in extensive form is Rosenthal's centipede game [53]. This game is interesting because the equilibrium identified by backward induction is almost never played in practical tests with humans. This type of investigation falls under the area of behavioral game theory or behavioral economics [54].

In addition to its use in economics, game theory is also used in political science. The Battle of the Bismark Sea example comes from Brams' book, *Game Theory and Politics* [40]. Game theory has been extensively used in military analysis with varying levels of success (see, e.g., Refs. [55–58]).

$$- \spadesuit \clubsuit \heartsuit \diamondsuit -$$

3.7   Exercises

3.1   Compute the number of directed graphs on four vertices. [Hint: How many different pairs of vertices are there?]

3.2   Using the approach from Example 3.31, derive a strategy for Player 2 in the rock-paper-scissors game (Example 3.27) assuming she will attempt to maximize her payoff. Similarly, show that it doesn't matter whether Player 1 chooses rock, paper, or scissors in this game, and thus any strategy for Player 1 is equally good (or bad).

3.3   Consider a simplified game of tic-tac-toe where the objective is to fill in a board shown in Fig. 3.16.

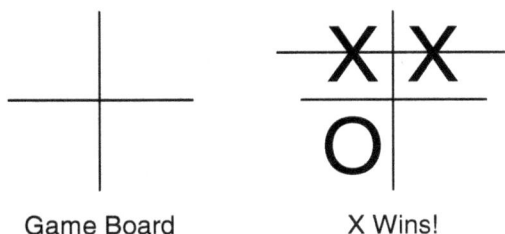

Game Board   X Wins!

Fig. 3.16. Players in this game try to get two in a row.

Assume that X goes first. Construct the game tree for this game by assuming that the winner receives $+1$ while the loser receives -1, and draws result in 0 for both players. Compute the depth of the longest path in the game tree. Show that there is a strategy so that the first player always wins. [Hint: You will need to consider each position on the board as one of the moves that can be made.]

3.4 On a standard 3×3 tic-tac-toe board, compute the length of the longest path in the game tree. [Hint: Assume you draw in this game.]

3.5 Consider the information sets as a collection of labels $\mathcal{I}$, and let $\xi : V \to \mathcal{I}$. Write down the constraints that ξ must satisfy so that this definition of information set is analogous to Definition 3.35.

3.6 Identify the information sets for a regular game of rock-paper-scissors, and draw the game tree to illustrate the incomplete information. You do not need to identify an optimal strategy for either player.

3.7 Define a strategy for rock-paper-scissors, and show the unique path through the tree in Fig. 3.5 determined by this strategy. Do the same for the game tree describing the Battle of the Bismark Sea with incomplete information.

3.8   Draw a game tree for the following game: At the beginning of this game, each player antes up $1 into a common pot. Player 1 takes a card from a randomized (shuffled) deck. After looking at the card, Player 1 decides whether to raise or fold:

(1) If Player 1 folds, he shows the card to Player 2; if the card is red, then Player 1 wins the pot and Player 2 loses the pot, whereas if the card is black, then Player 1 loses the pot and Player 2 wins the pot.
(2) If Player 1 raises, then Player 1 adds another dollar to the pot, and Player 2 picks a card and must decide whether to call or fold.

    (a) If Player 2 folds, then the game ends, and Player 1 takes the money, irrespective of any cards drawn.
    (b) If Player 2 calls, then he adds $1 to the pot. Both players show their cards; if both cards are of the same suit, then Player 1 wins the pot ($2) and Player 2 loses the pot, whereas if the cards are of opposite suits, then Player 2 wins the pot and Player 1 loses.

3.9   Continuing from Exercise 3.8, draw the game tree when we know that Player 1 is dealt a red card. Illustrate in your drawing how it is a subtree of the tree you drew in Exercise 3.8. Determine whether this game is still (i) a game of chance and (ii) whether it is a complete-information game or not.

3.10   Suppose that players always raise and call in the game defined in Exercise 3.8. Compute the probability space defined by these strategies in the game tree you developed.

3.11   Decide whether the strategy (Raise, Call) is an equilibrium strategy in the game in Example 3.49.

3.12 Show that in rock-paper-scissors with perfect information, there are three equilibrium strategies.

3.13 Prove Zermelo's theorem. Can you illustrate a game of this type? [Hint: Use Theorems 3.67 and 3.53. There are many games of this type.]

Chapter 4

Games and Matrices: Normal and Strategic Forms

Chapter Goals: The goal of this chapter is to introduce games in normal and strategic forms. Games in strategic form are usually called matrix games. We also discuss strategy vectors and show how payoffs can be computed using simple matrix arithmetic. Equilibria are defined in terms of strategy vectors. This chapter forms the foundation for results in later chapters. An introduction to the elements of matrix arithmetic needed to understand this chapter is provided in Appendix A.

4.1   Normal and Strategic Forms

Definition 4.1 (Normal Form). A game in *normal form* is a triple $\mathcal{G} = (\mathbf{P}, \Sigma, \pi)$, where $\mathbf{P}$ is the player set, $\Sigma = \Sigma_1 \times \Sigma_2 \times \cdots \times \Sigma_N$ is a discrete strategy space, and $\pi : \Sigma \to \mathbb{R}^N$ is a strategy payoff function.

Remark 4.2. If $\mathcal{G} = (\mathbf{P}, \Sigma, \pi)$ is a normal-form game, then the function $\pi_i : \Sigma \to \mathbb{R}$ is the payoff function for Player P_i and returns the ith component of the function π.

Remark 4.3. The notation used in Definition 4.1 is identical to that introduced in Chapter 3. Consequently, it is straightforward to see that a game in extensive form can be converted into a game in normal form.

Definition 4.4 (Constant/General-Sum Game). Let $\mathcal{G} = (\mathbf{P}, \Sigma, \pi)$ be a game in normal form. If there is a constant $C \in \mathbb{R}$ so that for all strategy tuples $(\sigma_1, \ldots, \sigma_N) \in \Sigma$, we have

$$\sum_{i=1}^{N} \pi_i(\sigma_1, \ldots, \sigma_N) = C, \tag{4.1}$$

then $\mathcal{G}$ is called a *constant-sum game*. If $C = 0$, then $\mathcal{G}$ is called a *zero-sum game*. Any game that is *not* constant sum is called a *general-sum game*.

Example 4.5. This example is a variation on the one by Brian Burke [59] on his blog. A North American Football play (in which the score does not change) is an example of a zero-sum game when the payoff is measured by yards gained or lost. In this game, there are two players: the Offense (P_1) and the Defense (P_2). The Offense may choose between two strategies:

$$\Sigma_1 = \{\text{Pass}, \text{Run}\}. \tag{4.2}$$

The Defense may choose between three strategies:

$$\Sigma_2 = \{\text{Pass Defense}, \text{Run Defense}, \text{Blitz}\}. \tag{4.3}$$

The yards gained by the Offense are lost by the Defense. Suppose the following payoff function (in terms of yards gained or lost by each player) π is defined:

$$\pi(\text{Pass}, \text{Pass Defense}) = (-2, 2),$$
$$\pi(\text{Pass}, \text{Run Defense}) = (8, -8),$$
$$\pi(\text{Pass}, \text{Blitz}) = (-4, 4),$$
$$\pi(\text{Run}, \text{Pass Defense}) = (6, -6),$$
$$\pi(\text{Run}, \text{Run Defense}) = (-2, 2),$$
$$\pi(\text{Run}, \text{Blitz}) = (5, -5).$$

If $\mathbf{P} = \{P_1, P_2\}$ and $\Sigma = \Sigma_1 \times \Sigma_2$, then the tuple $\mathcal{G} = (\mathbf{P}, \Sigma, \pi)$ is a zero-sum game in normal form. Note that each pair in the definition of the payoff function sums to zero.

Remark 4.6. Just as in a game in extensive form, we can define an equilibrium. This definition is identical to the definition we gave in Definition 3.63.

Definition 4.7 (Equilibrium). A strategy $(\sigma_1^*, \ldots, \sigma_N^*) \in \Sigma$ is an equilibrium if, for all i,

$$\pi_i(\sigma_1^*, \ldots, \sigma_i^*, \ldots, \sigma_N^*) \geq \pi_i(\sigma_1^*, \ldots, \sigma_i, \ldots, \sigma_N^*),$$

where $\sigma_i \in \Sigma_i$.

4.2 Strategic-Form Games

Remark 4.8. For the remainder of this book, we assume that the reader is familiar with matrix arithmetic. A review of all the necessary facts can be found in Appendix A. It is worth noting that we use some notation common to operations research for convenience. The ith row of a matrix $\mathbf{A} \in \mathbb{R}^{m \times n}$ (i.e., an $m \times n$ rectangular array of real numbers) is denoted $A_{i \cdot}$, while the jth column of that matrix is denoted $\mathbf{A}_{\cdot j}$. This notation appears throughout the remainder of the book.

Definition 4.9 (Strategic Form – Two-Player Games). Let $\mathcal{G} = (\mathbf{P}, \Sigma, \pi)$ be a normal-form game, with $\mathbf{P} = \{P_1, P_2\}$ and $\Sigma = \Sigma_1 \times \Sigma_2$. Suppose the strategies in Σ_i $(i = 1, 2)$ are ordered so that $\Sigma_i = \{\sigma_1^i, \ldots, \sigma_{n_i}^i\}$ $(i = 1, 2)$. Then, there are two matrices $\mathbf{A}, \mathbf{B} \in \mathbb{R}^{n_1 \times n_2}$ so that

$$\mathbf{A}_{rc} = \pi_1(\sigma_r^1, \sigma_c^2),$$
$$\mathbf{B}_{rc} = \pi_2(\sigma_r^1, \sigma_c^2).$$

That is, the (r, c) element of the matrices are given by the payoff functions. Then, the tuple $\mathcal{G} = (\mathbf{P}, \Sigma, \mathbf{A}, \mathbf{B})$ is a *two-player game in strategic form*.

Remark 4.10. Games with two players given in strategic form are also sometimes called *matrix games* or *bimatrix games* because they are defined completely by matrices. Note also that, by convention, Player P_1's strategies correspond to the rows of the matrices, while Player P_2's strategies correspond to the columns of the matrices.

Example 4.11. Consider the two-player game defined in the Battle of the Bismark Sea. If we assume that the strategies for the players are

$$\Sigma_1 = \{\text{Sail North}, \text{Sail South}\},$$
$$\Sigma_2 = \{\text{Search North}, \text{Search South}\},$$

then the payoff matrices for the two players are

$$\mathbf{A} = \begin{bmatrix} -2 & -1 \\ -2 & -3 \end{bmatrix},$$

$$\mathbf{B} = \begin{bmatrix} 2 & 1 \\ 2 & 3 \end{bmatrix}.$$

Here, the *rows* represent the different strategies of Player 1, and the *columns* represent the strategies of Player 2. Thus, the $(1,1)$ entry in the matrix $\mathbf{A}$ is the payoff to Player 1 when the strategy pair (Sail North, Search North) is played. The $(2,1)$ entry in matrix $\mathbf{B}$ is the payoff to Player 2 when the strategy pair (Sail South, Search North) is played, etc. Note in this case that $\mathbf{A} = -\mathbf{B}$. This is because the Battle of the Bismark Sea is a zero-sum game.

Example 4.12 (Chicken). Consider the following two-player game: Two cars face each other and begin driving (quickly) toward each other (see Fig. 4.1.). The player who swerves first loses 1 point, while the other player wins 1 point. If both players swerve, then each receives 0 points. If neither player swerves, a bad crash occurs and both players lose 10 points. (In reality, a crash is worse than losing points, but we must assign a numeric value.)

Assuming that the strategies for Player 1 are in the rows and the strategies for Player 2 are in the columns, then the two matrices for the players are

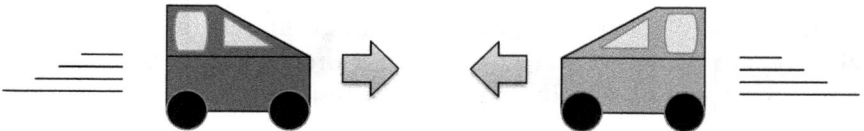

Fig. 4.1.   Illustration of a game of Chicken.

	Swerve	Don't Swerve
Player 1		
Swerve	0	−1
Don't Swerve	1	−10
Player 2		
Swerve	0	1
Don't Swerve	−1	−10

From this, we can see that the matrices are

$$\mathbf{A} = \begin{bmatrix} 0 & -1 \\ 1 & -10 \end{bmatrix},$$

$$\mathbf{B} = \begin{bmatrix} 0 & 1 \\ -1 & -10 \end{bmatrix}.$$

Note that Chicken is **not** a zero-sum game; it is a general-sum game.

Remark 4.13. Chicken (sometimes called Hawk-Dove or Snowdrift) can be generalized in the sense that we could write the payoff matrix for Player 1 as

$$\mathbf{A} = \begin{bmatrix} T & L \\ W & X \end{bmatrix},$$

where $W > T > L > X$ are arbitrary values. As expected, the payoff matrix for Player 2 is then $\mathbf{B} = \mathbf{A}^T$. For our examples, we use the numerical payoff matrices and leave generalization to the reader as appropriate.

Remark 4.14. Definition 4.9 can be extended to N-player games. However, we no longer have matrices with payoff values for various strategies. Instead, we construct N N-dimensional arrays. So, a game with 3 players yields 3 arrays with dimension 3. This is illustrated in Fig. 4.2.

Multidimensional arrays are easy to represent on computers but difficult to represent on paper. They have multiple indices instead of just one index like a vector or two indices like a matrix. The elements of the array for Player i store the various payoffs for Player i under different strategic combinations of the different players. If there are

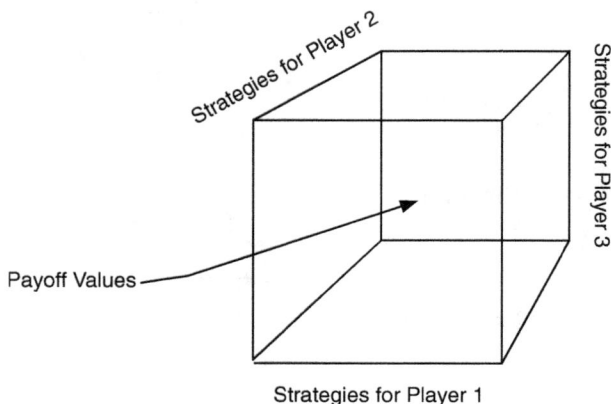

Fig. 4.2. A three-dimensional array is like a matrix with an extra dimension. They are difficult to capture on a page. The elements of the array for Player i store the various payoffs for Player i under different strategic combinations of the different players. If there are three players, then there will be three different arrays.

three players, then there will be three different arrays, one for each player.

Remark 4.15. The normal form of a (two-player) game is essentially the recipe for transforming a game in extensive form into a game in strategic form. Any game in extensive form can be transformed in this way, and the strategic form can be analyzed. Reasons for doing this include the fact that the strategic form is substantially more compact. However, it can be complex to compute if the size of the game tree in extensive form is very large.

Definition 4.16 (Zero-Sum Matrix Game). Suppose $\mathcal{G} = (\mathbf{P}, \Sigma, \mathbf{A}, \mathbf{B})$ is a game in strategic form and $\mathbf{B} = -\mathbf{A}$, so the game is zero sum. Then, we denote such a game as $\mathcal{G} = (\mathbf{P}, \Sigma, \mathbf{A})$, because the information in $\mathbf{B}$ is redundant.

Definition 4.17 (Symmetric Game). Let $\mathcal{G} = (\mathbf{P}, \Sigma, \mathbf{A}, \mathbf{B})$. If $\mathbf{A} = \mathbf{B}^T$, then $\mathcal{G}$ is called a *symmetric game*.

Example 4.18. Chicken is an example of a symmetric game.

4.3 Strategy Vectors and Matrix Games

Remark 4.19. Our next proposition relates the strategy set Σ to pairs of standard basis vectors and reduces the payoff function computation to simple matrix multiplication. Recall that

$$\mathbf{e}_i = \langle \underbrace{0, 0, \ldots, 0}_{i-1}, 1, \underbrace{0, 0, \ldots, 0}_{n-i} \rangle$$

is the ith standard basis vector, as discussed in Appendix A.

Proposition 4.20. *Let* $\mathcal{G} = (\mathbf{P}, \Sigma, \mathbf{A}, \mathbf{B})$ *be a two-player game in strategic form with* $\Sigma_1 = \{\sigma_1^1, \ldots, \sigma_m^1\}$ *and* $\Sigma_2 = \{\sigma_1^2, \ldots, \sigma_n^2\}$. *If Player* P_1 *chooses strategy* σ_r^1 *and Player* P_2 *chooses strategy* σ_c^2, *then*

$$\pi_1(\sigma_r^1, \sigma_c^2) = \mathbf{e}_r^T \mathbf{A} \mathbf{e}_c, \tag{4.4}$$

$$\pi_2(\sigma_r^1, \sigma_c^2) = \mathbf{e}_r^T \mathbf{B} \mathbf{e}_c. \tag{4.5}$$

Proof. For any matrix $\mathbf{A} \in \mathbb{R}^{m \times n}$, $\mathbf{A} \mathbf{e}_c$ returns column c of matrix $\mathbf{A}$, that is, $\mathbf{A}_{\cdot c}$. Likewise, $\mathbf{e}_r^T \mathbf{A}_{\cdot c}$ is the rth element of this vector. Thus, $\mathbf{e}_r^T \mathbf{A} \mathbf{e}_c$ is the (r, c)th element of the matrix $\mathbf{A}$. By definition, this must be the payoff for the strategy pair (σ_r^1, σ_c^2) for Player P_1. A similar argument follows for Player P_2 and matrix $\mathbf{B}$. □

Remark 4.21. Proposition 4.20 says that for two-player matrix games, we can relate any choice of strategy that Player P_i makes with a unit vector. Thus, we can define the payoff function in terms of vector and matrix multiplication. We will see that this can be generalized to cases when the strategies of the players are *not* represented by standard basis vectors.

Example 4.22. Consider the game of Chicken. Suppose Player P_1 decides to swerve, while Player P_2 decides not to swerve. Then, we can represent the strategy of Player P_1 by the vector

$$\mathbf{e}_1 = \begin{bmatrix} 1 \\ 0 \end{bmatrix},$$

while the strategy of Player P_2 is represented by the vector

$$\mathbf{e}_2 = \begin{bmatrix} 0 \\ 1 \end{bmatrix}.$$

Recall that the payoff matrices for this game are

$$\mathbf{A} = \begin{bmatrix} 0 & -1 \\ 1 & -10 \end{bmatrix},$$

$$\mathbf{B} = \begin{bmatrix} 0 & 1 \\ -1 & -10 \end{bmatrix}.$$

Then, we can compute

$$\pi_1(\text{Swerve}, \text{Don't Swerve}) = \mathbf{e}_1^T \mathbf{A} \mathbf{e}_2 = \begin{bmatrix} 1 & 0 \end{bmatrix} \cdot \begin{bmatrix} 0 & -1 \\ 1 & -10 \end{bmatrix} \cdot \begin{bmatrix} 0 \\ 1 \end{bmatrix} = -1,$$

$$\pi_2(\text{Swerve}, \text{Don't Swerve}) = \mathbf{e}_1^T \mathbf{B} \mathbf{e}_2 = \begin{bmatrix} 1 & 0 \end{bmatrix} \cdot \begin{bmatrix} 0 & 1 \\ -1 & -10 \end{bmatrix} \cdot \begin{bmatrix} 0 \\ 1 \end{bmatrix} = 1.$$

We can also consider the case when both players swerve. Then, we can represent the strategies of both Players by $\mathbf{e}_1$. In this case, we have

$$\pi_1(\text{Swerve}, \text{Swerve}) = \mathbf{e}_1^T \mathbf{A} \mathbf{e}_1 = \begin{bmatrix} 1 & 0 \end{bmatrix} \cdot \begin{bmatrix} 0 & -1 \\ 1 & -10 \end{bmatrix} \cdot \begin{bmatrix} 1 \\ 0 \end{bmatrix} = 0,$$

$$\pi_2(\text{Swerve}, \text{Swerve}) = \mathbf{e}_1^T \mathbf{B} \mathbf{e}_1 = \begin{bmatrix} 1 & 0 \end{bmatrix} \cdot \begin{bmatrix} 0 & 1 \\ -1 & -10 \end{bmatrix} \cdot \begin{bmatrix} 1 \\ 0 \end{bmatrix} = 0.$$

Remark 4.23. We now define equilibria in terms of matrix operations.

Proposition 4.24 (Equilibrium). *Let $\mathcal{G} = (\mathbf{P}, \Sigma, \mathbf{A}, \mathbf{B})$ be a two-player game in strategic form with $\Sigma = \Sigma_1 \times \Sigma_2$. The expressions*

$$\mathbf{e}_i^T \mathbf{A} \mathbf{e}_j \geq \mathbf{e}_k^T \mathbf{A} \mathbf{e}_j \quad \forall k \neq i \tag{4.6}$$

and

$$\mathbf{e}_i^T \mathbf{B} \mathbf{e}_j \geq \mathbf{e}_i^T \mathbf{B} \mathbf{e}_l \quad \forall l \neq j \tag{4.7}$$

hold if and only if $(\sigma_i^1, \sigma_j^2) \in \Sigma_1 \times \Sigma_2$ is an equilibrium strategy.

Proof. From Proposition 4.20, we know that

$$\pi_1(\sigma_i^1, \sigma_j^2) = \mathbf{e}_i^T \mathbf{A} \mathbf{e}_j, \tag{4.8}$$

$$\pi_2(\sigma_i^1, \sigma_j^2) = \mathbf{e}_i^T \mathbf{B} \mathbf{e}_j. \tag{4.9}$$

From Eq. (4.6), we know that for all $k \neq i$,

$$\pi_1(\sigma_i^1, \sigma_j^2) \geq \pi_1(\sigma_k^1, \sigma_j^2). \tag{4.10}$$

From Eq. (4.7), we know that for all $l \neq j$,

$$\pi_2(\sigma_i^1, \sigma_j^2) \geq \pi_2(\sigma_i^1, \sigma_l^2). \tag{4.11}$$

Thus, from Definition 4.7, it is clear that $(\sigma_i^1, \sigma_j^2) \in \Sigma$ is an equilibrium strategy. The converse is clear from this as well. □

Remark 4.25. We can now think of relating a strategy choice for player i, $\sigma_k^i \in \Sigma_i$, with the unit vector $\mathbf{e}_k$. From context, we will be able to identify to which player's strategy vector $\mathbf{e}_k$ corresponds.

4.4 Chapter Notes

Matrices (in the form used in this chapter) are a relatively recent invention, though arrays of numbers have been used for centuries. The term matrix was first used by J. J. Sylvester in Ref. [60]. At this time, matrices were not used for computational simplicity, as they are today. In the following chapters, we will see how the matrix representation of games can simplify our analysis.

Some texts list both the matrices $\mathbf{A}$ and $\mathbf{B}$ together. In this form, for example, Chicken is described by the table

$$\begin{bmatrix} (0,0) & (-1,1) \\ (1,-1) & (-10,-10) \end{bmatrix}.$$

This can be compact but has the potential to cause confusion. As a rule, the row player is always player one, while the column player is always player two.

Interestingly, Chicken formed the basis for the early American and Soviet nuclear policies under the concept of Mutual Assured Destruction (MAD) [27]. In the MAD policy, the idea was to build in an automated nuclear response to any attack, thus ensuring that "Don't Swerve" (i.e., attack) was played by both players (the United States and the Soviet Union) automatically. In this way, no side acting rationally would ever execute a first strike because the result would be catastrophic.

$$- \spadesuit \clubsuit \heartsuit \diamondsuit -$$

4.5   Exercises

4.1   Compute the payoff matrices for Example 4.5.

4.2   Construct payoff matrices for rock-paper-scissors. Also, construct the normal form of the game.

4.3   Compute the strategic form of the two-player coin-flipping game using the expected payoff function defined in Example 3.62.

4.4   Confirm that $(\mathbf{e}_1, \mathbf{e}_2)$ is an equilibrium strategy for the game of Chicken using the numeric payoff matrices given in Example 4.12. Does this generalize to the case with arbitrary values W, T, L, and X, as in Remark 4.13? Use symmetry to find a second pure strategy equilibrium.

Chapter 5

Saddle Points, Mixed Strategies, and Nash Equilibria

Chapter Goals: In this chapter, we focus on equilibria in games. We introduce the concept of a Nash equilibrium, which fully generalizes the previous notions of equilibria we have discussed. We study dominated strategies and their relationship with Nash equilibria. We then prove the minimax theorem and study Nash's original proofs of the existence of Nash equilibria. The results in this chapter are used as a basis for the remainder of the book, which uses techniques from optimization to find equilibria.

Remark 5.1 (Notational Remark). For the remainder of the book, unless otherwise noted, we assume that $\mathcal{G} = (\mathbf{P}, \Sigma, \pi)$ is a game in normal form, with $\mathbf{P} = \{P_1, \ldots, P_N\}$ and $\Sigma_i = \{\sigma_1^i, \ldots, \sigma_{n_i}^i\}$. When we discuss two-player games, we assume that $\Sigma = \Sigma_1 \times \Sigma_2$, $\Sigma_1 = \{\sigma_1^1, \ldots, \sigma_m^1\}$, and $\Sigma_2 = \{\sigma_1^2, \ldots, \sigma_n^2\}$. Therefore, a zero-sum game is the tuple $\mathcal{G} = (\mathbf{P}, \Sigma, \mathbf{A})$, with $\mathbf{A} \in \mathbb{R}^{m \times n}$, while a bimatrix game is the tuple $\mathcal{G} = (\mathbf{P}, \Sigma, \mathbf{A}, \mathbf{B})$, with $\mathbf{A}, \mathbf{B} \in \mathbb{R}^{m \times n}$.

5.1   Equilibria in Zero-Sum Games: Saddle Points

Theorem 5.2. *Let* $\mathcal{G} = (\mathbf{P}, \Sigma, \mathbf{A})$ *be a zero-sum two-player game. A strategy pair,* $(\mathbf{e}_i, \mathbf{e}_j)$, *is an equilibrium strategy if and only if*

$$\mathbf{e}_i^T \mathbf{A} \mathbf{e}_j = \max_{k \in \{1, \ldots, m\}} \min_{l \in \{1, \ldots, n\}} \mathbf{A}_{kl} = \min_{l \in \{1, \ldots, n\}} \max_{k \in \{1, \ldots, m\}} \mathbf{A}_{kl}. \quad (5.1)$$

Example 5.3. Before we prove Theorem 5.2, let's first consider an example from television before streaming. Two network corporations believe that there are $100M$ viewers to be had during Thursday night prime time (8–9 pm). The corporations must decide which type of programming to run: science fiction, drama, or comedy. If the two networks initially split the $100M$ viewers evenly, we can think of the payoff matrix as determining how many excess viewers the networks' strategies will yield over $50M$. The payoff matrix (in millions) for Network 1 is shown in Eq. (5.2):

$$\mathbf{A} = \begin{bmatrix} -14 & -34 & 11 \\ -4 & 9 & 0 \\ -11 & -35 & 21 \end{bmatrix}. \quad (5.2)$$

That is, if Network 1 and Network 2 both choose strategy one, then Network 1 has $36 = 50 - 14$ million viewers and Network 2 has $64 = 50 + 14$ million viewers. The expression

$$\min_{l \in \{1, \ldots, n\}} \max_{k \in \{1, \ldots, m\}} \mathbf{A}_{kl}$$

asks us to compute the maximum value in each column to create the set

$$C_{\max} = \{c_l^* = \max\{\mathbf{A}_{kl} : k \in \{1, \ldots, m\}\} : l \in \{1, \ldots, n\}\}$$

and then choose the smallest value in this case. If we look at this matrix, the column maximums are

$$\begin{bmatrix} -4 & 9 & 21 \end{bmatrix}.$$

We then choose the minimum value in this case, which is -4. This value occurs at position $(2, 1)$.

The expression

$$\max_{k \in \{1, \ldots, m\}} \min_{l \in \{1, \ldots, n\}} \mathbf{A}_{kl}$$

asks us to compute the minimum value in each row to create the set

$$R_{\min} = \{r_k^* = \min\{\mathbf{A}_{kl} \colon l \in \{1, \ldots, n\}\} \colon k \in \{1, \ldots, m\}\}$$

and then choose the largest value in this case. Again, if we look at the matrix in Eq. (5.2), we see that the minimum values in the rows are

$$\begin{bmatrix} -34 \\ -4 \\ -35 \end{bmatrix}.$$

The largest value in this case is -4. Again, this value occurs at position $(2, 1)$. This process is captured in Fig. 5.1. The two values are equal, and so by Theorem 5.63, the equilibrium strategy pair is $(\mathbf{e}_2, \mathbf{e}_1)$, which returns the value in the second row and first column to Network 1. Network 2 receives the negative of this value.

Remark 5.4. To understand why this works, consider the following logic. The row player (Player 1) knows that Player 2 (the column player) is trying to maximize her (Player 2's) payoff. Since this is a zero-sum game, any increase in Player 2's payoff will come at the expense of Player 1. So, Player 1 looks at each row independently (since he chooses rows) and asks, "What is the worst possible outcome I could see if I played a strategy corresponding to this row?" Having obtained these worst possible scenarios, he chooses the row with the highest value.

Payoff Matrix			Row Min
−14	−34	11	−34
−4	9	0	−4
−11	−35	21	−35
−4	8	20	maxmin = −4
Column Max			minmax = −4

Fig. 5.1. The minimax analysis of the game of competing networks. The row player knows that Player 2 (the column player) is trying to maximize her (Player 2's) payoff. Thus, Player 1 asks: "What is the worst possible outcome I could see if I played a strategy corresponding to this row?" Having obtained these *worst possible scenarios*, he chooses the row with the highest value. Player 2 does something similar in columns.

Player 2 faces a similar problem. She knows that Player 1 wishes to maximize his payoff and that any gain will come at her expense. So, Player 2 looks across each column of the matrix $\mathbf{A}$ and asks, "What is the best possible score Player 1 can achieve if I (Player 2) choose to play the strategy corresponding to the given column?" Remember that the negation of this value will be Player 2's payoff in this case. Having done that, Player 2 then chooses the column that minimizes this value and thus maximizes her payoff. If these two values are equal, then the theorem claims that the resulting strategy pair is an equilibrium.

Remark 5.5. We are now ready to prove Theorem 5.2.

Proof of Theorem 5.2. ($\Rightarrow$) Suppose that $(\mathbf{e}_i, \mathbf{e}_j)$ is an equilibrium solution. Then, we know that

$$\mathbf{e}_i^T \mathbf{A} \mathbf{e}_j \geq \mathbf{e}_k^T \mathbf{A} \mathbf{e}_j,$$

$$\mathbf{e}_i^T (-\mathbf{A}) \mathbf{e}_j \geq \mathbf{e}_i^T (-\mathbf{A}) \mathbf{e}_l,$$

for all $k \in \{1, \ldots, m\}$ and $l \in \{1, \ldots, n\}$. We can write this as

$$\mathbf{e}_i^T \mathbf{A} \mathbf{e}_j \geq \mathbf{e}_k^T \mathbf{A} \mathbf{e}_j \tag{5.3}$$

and

$$\mathbf{e}_i^T \mathbf{A} \mathbf{e}_j \leq \mathbf{e}_i^T \mathbf{A} \mathbf{e}_l. \tag{5.4}$$

We know that $\mathbf{e}_i^T \mathbf{A} \mathbf{e}_j = \mathbf{A}_{ij}$ and that Eq. (5.3) holds if and only if

$$\mathbf{A}_{ij} \geq \mathbf{A}_{kj}, \tag{5.5}$$

for all $k \in \{1, \ldots, m\}$. That is, the element i must be maximal in the column $\mathbf{A}_{\cdot j}$. Note that for a fixed row $k \in \{1, \ldots, m\}$,

$$\mathbf{A}_{kj} \geq \min\{\mathbf{A}_{kl} : l \in \{1, \ldots, n\}\}.$$

This means that if we compute the minimum value in a row k, then the value in column j, $\mathbf{A}_{kj}$, must be at least as large as that minimal value. Combining this with Eq. (5.5), we conclude that for each row $k \in \{1, \ldots, m\}$,

$$\mathbf{A}_{ij} \geq \min\{\mathbf{A}_{kl} : l \in \{1, \ldots, n\}\}. \tag{5.6}$$

However, Eq. (5.6) implies that

$$\mathbf{e}_i^T \mathbf{A} \mathbf{e}_j = \mathbf{A}_{ij} = \max_{k \in \{1,\ldots,m\}} \min_{l \in \{1,\ldots,n\}} A_{kl}. \tag{5.7}$$

Likewise, Eq. (5.4) holds if and only if

$$\mathbf{A}_{ij} \leq \mathbf{A}_{il}, \tag{5.8}$$

for all $l \in \{1,\ldots,n\}$. Arguing as before, for a fixed column $l \in \{1,\ldots,n\}$, we have

$$\mathbf{A}_{il} \leq \max\{\mathbf{A}_{kl}: k \in \{1,\ldots,m\}\}.$$

This means that if we compute the maximum value in a column l, then the value in row i, $\mathbf{A}_{il}$, must not exceed that maximal value. Combining this with Eq. (5.8) gives

$$\mathbf{A}_{ij} \leq \max\{\mathbf{A}_{kl}: k \in \{1,\ldots,m\}\}. \tag{5.9}$$

However, Eq. (5.9) implies that

$$\mathbf{e}_i^T \mathbf{A} \mathbf{e}_j = \mathbf{A}_{ij} = \min_{l \in \{1,\ldots,n\}} \max_{k \in \{1,\ldots,m\}} \mathbf{A}_{kl}. \tag{5.10}$$

Thus, it follows that

$$\mathbf{A}_{ij} = \mathbf{e}_i^T \mathbf{A} \mathbf{e}_j = \max_{k \in \{1,\ldots,m\}} \min_{l \in \{1,\ldots,n\}} \mathbf{A}_{ij} = \min_{l \in \{1,\ldots,n\}} \max_{k \in \{1,\ldots,m\}} \mathbf{A}_{kl}.$$

($\Leftarrow$) To prove the converse, suppose that

$$\mathbf{e}_i^T \mathbf{A} \mathbf{e}_j = \max_{k \in \{1,\ldots,m\}} \min_{l \in \{1,\ldots,n\}} \mathbf{A}_{kl} = \min_{l \in \{1,\ldots,n\}} \max_{k \in \{1,\ldots,m\}} \mathbf{A}_{kl}.$$

Consider the quantity

$$\mathbf{e}_k^T \mathbf{A} \mathbf{e}_j = \mathbf{A}_{kj}.$$

The fact that

$$\mathbf{A}_{ij} = \max_{k \in \{1,\ldots,m\}} \min_{l \in \{1,\ldots,n\}} \mathbf{A}_{kl}$$

implies that $\mathbf{A}_{ij} \geq \mathbf{A}_{kj}$ for any $k \in \{1,\ldots,m\}$. To see this, remember that

$$C_{\max} = \{c_l^* = \max\{\mathbf{A}_{kl}: k \in \{1,\ldots,m\}\} : l \in \{1,\ldots,n\}\} \tag{5.11}$$

and $\mathbf{A}_{ij} \in C_{\max}$ by construction. Thus, it follows that

$$\mathbf{e}_i^T \mathbf{A} \mathbf{e}_j \geq \mathbf{e}_k^T \mathbf{A} \mathbf{e}_j,$$

for any $k \in \{1, \ldots, m\}$. By a similar argument, we know that

$$\mathbf{A}_{ij} = \min_{l \in \{1,\ldots,m\}} \max_{k \in \{1,\ldots,n\}} \mathbf{A}_{kl},$$

which implies that $\mathbf{A}_{ij} \leq \mathbf{A}_{il}$ for any $l \in \{1, \ldots, n\}$. To see this, remember that

$$R_{\min} = \{r_k^* = \min\{\mathbf{A}_{kl} : l \in \{1, \ldots, n\}\} : k \in \{1, \ldots, m\}\}$$

and $\mathbf{A}_{ij} \in R_{\min}$ by construction. Thus, it follows that

$$\mathbf{e}_i^T \mathbf{A} \mathbf{e}_j \leq \mathbf{e}_i^T \mathbf{A} \mathbf{e}_l,$$

for any $l \in \{1, \ldots, n\}$. We conclude that $(\mathbf{e}_i, \mathbf{e}_j)$ is an equilibrium solution. This completes the proof. $\qquad\square$

Theorem 5.6. *Suppose that* $\mathcal{G} = (\mathbf{P}, \Sigma, \mathbf{A})$ *is a zero-sum two-player game. Let* $(\mathbf{e}_i, \mathbf{e}_j)$ *be an equilibrium strategy pair for this game. If* $(\mathbf{e}_k, \mathbf{e}_l)$ *is a second equilibrium strategy pair, then*

$$\mathbf{A}_{ij} = \mathbf{A}_{kl} = \mathbf{A}_{il} = \mathbf{A}_{kj}.$$

Definition 5.7 (Saddle Point). Let $\mathcal{G} = (\mathbf{P}, \Sigma, \mathbf{A})$ be a zero-sum two-player game. If $(\mathbf{e}_i, \mathbf{e}_j)$ is an equilibrium, then it is called a *saddle point*.

Definition 5.8 (Game Value). Let $\mathcal{G} = (\mathbf{P}, \Sigma, \mathbf{A})$ be a zero-sum game. If there exists a strategy pair $(\mathbf{e}_i, \mathbf{e}_j)$ so that

$$\max_{k \in \{1,\ldots,m\}} \min_{l \in \{1,\ldots,n\}} \mathbf{A}_{kl} = \min_{l \in \{1,\ldots,n\}} \max_{k \in \{1,\ldots,m\}} \mathbf{A}_{kl},$$

then

$$V_{\mathcal{G}} = \mathbf{e}_i^T \mathbf{A} \mathbf{e}_j \tag{5.12}$$

is the *value of the game*.

Remark 5.9. We see that we can define the value of a zero-sum game even when there is no equilibrium point in strategies in Σ. Using Theorem 5.6, we can see that this value is unique, that is, any equilibrium pair for a game will yield the same value for a zero-sum game. This is not the case in a general-sum game.

5.2   Zero-Sum Games without Saddle Points

Remark 5.10. Not all games have saddle points of the kind found in Example 5.3. The easiest way to show that this is true is to illustrate it with an example.

Example 5.11. In August 1944, after the invasion of Normandy, the Allies broke out of their beachhead at Avranches, France, and headed toward the main part of the country (see Fig. 5.2).[1] The German General von Kluge, commander of the ninth army, faced two options:

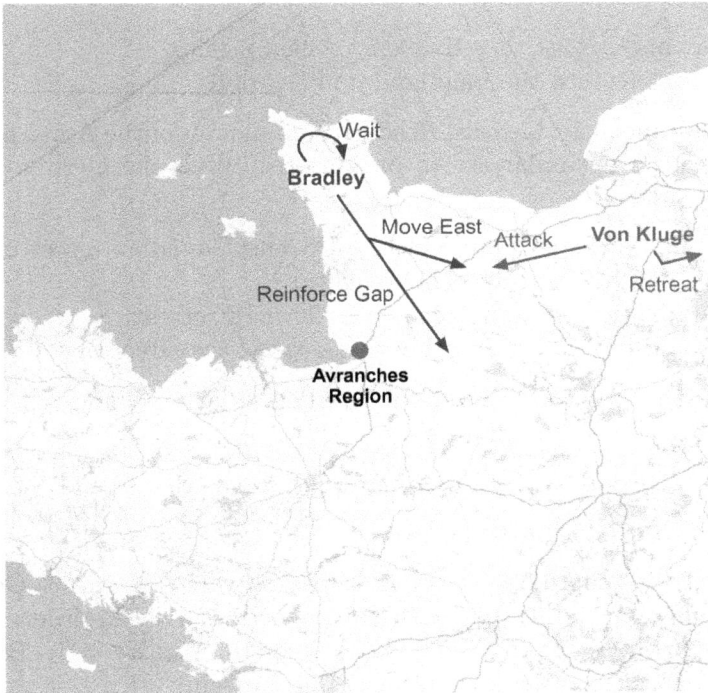

Fig. 5.2.   In August 1944, the allies broke out of their beachhead at Avranches. Each commander faced several troop movement choices. These choices can be modeled as a game. (Diagrammed troop movements are approximate.)

---

[1]This example is discussed in detail by Brams in Ref. [40].

Table 5.1. Game matrix of the battle of Avranches showing that this game has no saddle-point solution. There is no position in the matrix where an element is simultaneously the maximum value in its column and the minimum value in its row.

Bradley's Strategy	von Kluge's Strategies		Row Min
	Attack	Retreat	—
Reinforce Gap	2	3	2
Move East	1	5	1
Wait	6	4	4
Column Max	6	5	maxmin = 4
			minmax = 5

(1) Stay and attack the advancing Allied armies.

(2) Withdraw into the mainland and regroup.

Simultaneously, General Bradley, commander of the Allied ground forces, faced a similar set of options regarding the German ninth army:

(1) Reinforce the gap between the US and Canadian forces created by troop movements at Avranches.

(2) Send his forces east to cut off a German retreat.

(3) Do nothing and wait a day to see what the adversary did.

The player set **P** consists of Bradley (Player 1) and von Kluge (Player 2). The strategy sets are:

$$\Sigma_1 = \{\text{Reinforce the gap, Send forces east, Wait}\},$$

$$\Sigma_2 = \{\text{Attack, Retreat}\}.$$

In real life, there were no obvious pay-off values; however, General Bradley's diary indicates the scenarios he preferred in order. There are six possible scenarios, i.e., six elements in $\Sigma = \Sigma_1 \times \Sigma_2$. Bradley ordered them from most to least preferable, and using this ranking, we can construct the game matrix shown in Table 5.1.

Note that the maximin[2] value of the rows is not equal to the minimax value of the columns. This is indicative of the fact that

---

[2]The maxmin value is usually written maximin. Similarly, the minmax value is written minimax.

(Retreat, Wait)———————▷(Retreat, Move East)
(4, –4)   (5, –5)

(Attack, Wait)◁———————(Attack, Move East)
(6, –6)   (1, –1)

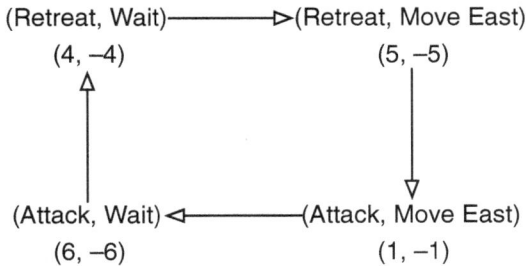

Fig. 5.3.   The Payoff values cause a cycle to occur in pure strategies.

there is not a pair of strategies that form an equilibrium for this game.

To illustrate this more clearly, suppose that von Kluge plays his minimax strategy to retreat. Then, Bradley would do better to not play his maximin strategy (wait) and instead move east, cutting off von Kluge's retreat, thus obtaining a payoff of $(5, -5)$. But von Kluge would realize this and deduce that he should attack, which would yield a payoff of $(1, -1)$. However, Bradley could deduce this as well and would know to play his maximin strategy (wait), which yields a payoff of $(6, -6)$. However, von Kluge would realize that this would occur, in which case he would decide to retreat, yielding a payoff of $(4, -4)$. The cycle then repeats. This logic is illustrated in Fig. 5.3.

5.3   Mixed Strategies

Remark 5.12. Heretofore, we have assumed that Player P_i will deterministically choose a strategy in Σ_i. It's possible, however, that Player P_i might choose a strategy at random. In this case, we assign a probability to each strategy in Σ_i.

Definition 5.13 (Mixed Strategy). A mixed strategy for Player $P_i \in \mathbf{P}$ is a discrete probability distribution function ρ^i defined over the sample space Σ_i. That is, we can define a discrete probability space $(\Sigma_i, \mathcal{F}_{\Sigma_i}, \rho^i)$ where Σ_i is the discrete sample space, $\mathcal{F}_{\Sigma_i}$ is the power set of Σ_i, and ρ^i is the discrete probability function that assigns probabilities to events in $\mathcal{F}_{\Sigma_i}$.

Table 5.2. The payoff matrix for Player P_1 in rock-paper-scissors. This payoff matrix can be derived from Fig. 3.5.

	Rock	**Paper**	**Scissors**
Rock	0	−1	1
Paper	1	0	−1
Scissors	−1	1	0

Remark 5.14. We assume that players choose their mixed strategies independently. Thus, we can compute the probability of a strategy element, $(\sigma^1, \ldots, \sigma^N) \in \Sigma$, as

$$\rho(\sigma^1, \ldots, \sigma^N) = \rho^1(\sigma^1)\rho^2(\sigma^2)\cdots\rho^N(\sigma^n). \tag{5.13}$$

Using this, we can define a discrete probability distribution over the sample space Σ as $(\Sigma, \mathcal{F}_\Sigma, \rho)$. Define Π_i as a random variable that maps Σ into $\mathbb{R}$ so that Π_i returns the payoff to Player P_i as a result of the random outcome $(\sigma^1, \ldots, \sigma^N)$. Therefore, the expected payoff for Player P_i for a given mixed strategy $(\rho^1, \ldots, \rho^N)$ is

$$\mathbb{E}(\Pi_i) = \sum_{\sigma^1 \in \Sigma_1} \sum_{\sigma^2 \in \Sigma_2} \cdots \sum_{\sigma^N \in \Sigma_N} \pi_i(\sigma^1, \ldots, \sigma^n)\rho^1(\sigma^1)\rho^2(\sigma^2)\cdots\rho^N(\sigma^N).$$

Example 5.15. Consider the rock-paper-scissors game. The payoff matrix for Player 1 is given in Table 5.2.

Suppose that each strategy is chosen with a probability of $\frac{1}{3}$ by each player. Then, the expected payoff to Player P_1 with this strategy is

$$\mathbb{E}(\Pi_1) = \left(\frac{1}{3}\right)\left(\frac{1}{3}\right)\pi_1(\text{Rock}, \text{Rock}) + \left(\frac{1}{3}\right)\left(\frac{1}{3}\right)\pi_1(\text{Rock}, \text{Paper})$$

$$+ \left(\frac{1}{3}\right)\left(\frac{1}{3}\right)\pi_1(\text{Rock}, \text{Scissors})$$

$$+ \left(\frac{1}{3}\right)\left(\frac{1}{3}\right)\pi_1(\text{Paper}, \text{Rock}) + \left(\frac{1}{3}\right)\left(\frac{1}{3}\right)\pi_1(\text{Paper}, \text{Paper})$$

$$+ \left(\frac{1}{3}\right)\left(\frac{1}{3}\right) \pi_1(\text{Paper}, \text{Scissors})$$

$$+ \left(\frac{1}{3}\right)\left(\frac{1}{3}\right) \pi_1(\text{Scissors}, \text{Rock})$$

$$+ \left(\frac{1}{3}\right)\left(\frac{1}{3}\right) \pi_1(\text{Scissors}, \text{Paper})$$

$$+ \left(\frac{1}{3}\right)\left(\frac{1}{3}\right) \pi_1(\text{Scissors}, \text{Scissors}) = 0.$$

We can likewise compute the same value for $\mathbb{E}(\pi_2)$ for Player P_2.

Remark 5.16. For the remainder of this book, we will be dealing with vectors of the form $\mathbf{x} = \langle x_1, \ldots, x_m \rangle \in \mathbb{R}^m$. In general, a bold lower-case symbol is a vector, and corresponding non-bold and indexed symbols are its entries. Vectors are discussed in Definition A.9 in Appendix A.

Definition 5.17 (Mixed-Strategy Vector). To any mixed strategy, ρ^i, for Player P_i, we may associate a *mixed-strategy vector*, $\mathbf{x}^i = \langle x_1^i, \ldots, x_{n_i}^i \rangle$, where

$$x_k^i = \rho^i\left(\sigma_k^i\right).$$

Remark 5.18. From Definition 5.17, we can deduce that if $\mathbf{x}^i$ is a mixed-strategy vector, then it must satisfy:

(1) $x_j^i \geq 0$ for $j \in \{1, \ldots, n_i\}$,
(2) $\sum_{j=1}^{n_i} x_j^i = 1$.

Moreover, these two properties are sufficient to ensure that we are defining a mathematically correct probability distribution over the strategy set Σ_i.

Definition 5.19 (Player Mixed-Strategy Space). The set

$$\Delta_{n_i} = \left\{ \langle x_1, \ldots, x_{n_i} \rangle \in \mathbb{R}^{n_i} : \sum_{i=1}^{n_i} x_i = 1; x_i \geq 0, i = 1, \ldots, n_i \right\} \tag{5.14}$$

is the *mixed-strategy space* in n_i dimensions for Player P_i.

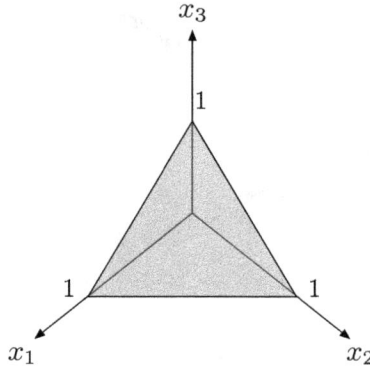

Fig. 5.4. In three-dimensional space, Δ_3 is the face of a tetrahedron. In four-dimensional space, it would be a tetrahedron, which would itself be the face of a four-dimensional object.

Remark 5.20. There is a pleasant geometry to the space Δ_n, which is usually called the *unit simplex* embedded in $\mathbb{R}^n$. In three dimensions, for example, the space is an equilateral triangle. (See Fig. 5.4.)

Remark 5.21. The space Δ_n is always an $n-1$-dimensional object that is embedded in (or lives in) $\mathbb{R}^n$. For this reason, some authors write Δ_{n-1} for Δ_n. Other authors use $\mathcal{S}_n$ or $\mathcal{S}_{n-1}$, where $\mathcal{S}$ denotes *simplex*. For pedagogic purposes, we have chosen to use Δ_n to remind the reader that there are n elements in a vector in Δ_n. More advanced texts may use different notation.

Definition 5.22 (Pure-Strategy Vector). The standard basis vector $\mathbf{e}_j \in \Delta_{n_i}$ corresponds to the strategy $\sigma_j^i \in \Sigma_i$ and, as such, is called a *pure-strategy vector*, or sometimes simply a *pure strategy*.

Definition 5.23 (Mixed-Strategy Space). The *mixed-strategy space* for the game $\mathcal{G}$ is the set

$$\Delta = \Delta_{n_1} \times \Delta_{n_2} \times \cdots \times \Delta_{n_N}. \tag{5.15}$$

Definition 5.24 (Mixed-Strategy Payoff Function). The expected payoff function, written in terms of the tuple of mixed-strategy vectors $(\mathbf{x}^1, \ldots, \mathbf{x}^N)$, is

$$u_i(\mathbf{x}^1, \ldots, \mathbf{x}^N) = \sum_{i_1=1}^{n_1} \sum_{i_2=1}^{n_2} \cdots \sum_{i_N=1}^{n_N} \pi_i(\sigma_{i_1}^1, \ldots, \sigma_{i_N}^n) \mathbf{x}_{i_1}^1 \mathbf{x}_{i_2}^2 \cdots \mathbf{x}_{i_N}^N.$$

$$\tag{5.16}$$

Here, $\mathbf{x}_i^j$ is the ith element of the vector $\mathbf{x}^j$. The function $u_i \colon \Delta \to \mathbb{R}$, defined in Eq. (5.16), is the *mixed-strategy payoff function* for Player P_i.

Example 5.25. For rock-paper-scissors, since each player has three strategies, $n = 3$ and Δ_3 consists of the vectors $\langle x_1, x_2, x_3 \rangle$ so that $x_1, x_2, x_3 \geq 0$ and $x_1 + x_2 + x_3 = 1$. For example, the vectors

$$\mathbf{x} = \mathbf{y} = \begin{bmatrix} \frac{1}{3} \\ \frac{1}{3} \\ \frac{1}{3} \end{bmatrix}$$

are mixed strategies for Players 1 and 2, respectively, that instruct the players to play rock 1/3 of the time, paper 1/3 of the time, and scissors 1/3 of the time.

Definition 5.26 (**Nash Equilibrium**). A *Nash equilibrium* is a tuple of mixed strategies, $(\mathbf{x}^{1*}, \ldots, \mathbf{x}^{N*}) \in \Delta$, so that for all $i \in \{1, \ldots, N\}$,

$$u_i(\mathbf{x}^{1*}, \ldots, \mathbf{x}^{i*}, \ldots, \mathbf{x}^{N*}) \geq u_i(\mathbf{x}^{1*}, \ldots, \mathbf{x}^{i}, \ldots, \mathbf{x}^{N*}), \qquad (5.17)$$

for all $\mathbf{x}^i \in \Delta_{n_i}$. If the inequality in the definition is strict, then it is a *strict* Nash equilibrium.

Remark 5.27. What Definition 5.26 states is that a tuple of mixed strategies, $(\mathbf{x}^{1*}, \ldots, \mathbf{x}^{N*})$, is a Nash equilibrium if *no player* has any reason to deviate *unilaterally* from her mixed strategy.

Remark 5.28 (**Notational Remark**). In many texts, it becomes cumbersome in N-player games to denote the mixed-strategy tuple $(\mathbf{x}^1, \ldots, \mathbf{x}^N)$, especially since we are usually only interested in the arbitrary Player P_i. To deal with this, textbooks sometimes adopt the notation $(\mathbf{x}^i, \mathbf{x}^{-i})$. Here, $\mathbf{x}^i$ is the mixed strategy for Player P_i, while $\mathbf{x}^{-i}$ denotes the mixed-strategy tuple for the other players (who are not Player P_i). When expressed this way, Eq. (5.17) is written as

$$u_i(\mathbf{x}^{i*}, \mathbf{x}^{-i*}) \geq u_i(\mathbf{x}^i, \mathbf{x}^{-i*}),$$

for all $i \in \{1, \ldots, N\}$. While notationally convenient, we restrict our attention to two-player games, so this will generally not be necessary.

Proposition 5.29. *Let $\mathcal{G} = (\mathbf{P}, \Sigma, \mathbf{A}, \mathbf{B})$ be a two-player matrix game. Let $\mathbf{x} \in \Delta_m$ and $\mathbf{y} \in \Delta_n$ be mixed strategies for Players 1 and 2, respectively. Then,*

$$u_1(\mathbf{x}, \mathbf{y}) = \mathbf{x}^T \mathbf{A} \mathbf{y}, \tag{5.18}$$

$$u_2(\mathbf{x}, \mathbf{y}) = \mathbf{x}^T \mathbf{B} \mathbf{y}. \tag{5.19}$$

Proof.   For simplicity, let $\mathbf{x} = \langle x_1, \ldots, x_m \rangle$ and $\mathbf{y} = \langle y_1, \ldots, y_n \rangle$. We know that $\pi_1(\sigma_i^1, \sigma_j^2) = \mathbf{A}_{ij}$. Simple matrix multiplication yields

$$\mathbf{x}^T \mathbf{A} = \begin{bmatrix} \mathbf{x}^T \mathbf{A}_{\cdot 1} \cdots \mathbf{x}^T \mathbf{A}_{\cdot n} \end{bmatrix}.$$

That is, $\mathbf{x}^T \mathbf{A}$ is a row vector whose jth element is $\mathbf{x}^T \mathbf{A}_{\cdot j}$. For fixed j, we have

$$\mathbf{x}^T \mathbf{A}_{\cdot j} = x_1 \mathbf{A}_{1j} + x_2 \mathbf{A}_{2j} + \cdots + x_m \mathbf{A}_{mj} = \sum_{i=1}^{m} \pi_1(\sigma_i^1, \sigma_j^2) x_i.$$

From this, we can conclude that

$$\mathbf{x}^T \mathbf{A} \mathbf{y} = \begin{bmatrix} \mathbf{x}^T \mathbf{A}_{\cdot 1} \cdots \mathbf{x}^T \mathbf{A}_{\cdot n} \end{bmatrix} \begin{bmatrix} y_1 \\ y_2 \\ \vdots \\ y_n \end{bmatrix}.$$

This simplifies to

$$\mathbf{x}^T \mathbf{A}_{\cdot 1} y_1 + \cdots + \mathbf{x}^T \mathbf{A}_{\cdot n} y_n$$

$$= (x_1 \mathbf{A}_{11} + x_2 \mathbf{A}_{21} + \cdots + x_m \mathbf{A}_{m1}) y_1$$

$$+ \cdots + (x_1 \mathbf{A}_{1n} + x_2 \mathbf{A}_{2n} + \cdots + x_m \mathbf{A}_{mn}) y_m. \tag{5.20}$$

Distributing the multiplication through, we can simplify Eq. (5.20) as

$$\mathbf{x}^T \mathbf{A} \mathbf{y} = \sum_{i=1}^{m} \sum_{j=1}^{n} \mathbf{A}_{ij} x_i y_j = \sum_{i=1}^{m} \sum_{j=1}^{n} \pi_1(\sigma_i^1, \sigma_j^2) x_i y_j = u_1(\mathbf{x}, \mathbf{y}). \tag{5.21}$$

A similar argument shows that $u_2(\mathbf{x}, \mathbf{y}) = \mathbf{x}^T \mathbf{B} \mathbf{y}$. This completes the proof. $\square$

5.4　Dominated Strategies and Nash Equilibria

Definition 5.30 (Weak Dominance). A mixed strategy, $\mathbf{x}^i \in \Delta_{n_i}$, for Player P_i *weakly dominates* another strategy, $\mathbf{y}^i \in \Delta_{n_i}$, for Player P_i if *for all* mixed strategies $\mathbf{z}^{-i}$, we have

$$u_i(\mathbf{x}^i, \mathbf{z}^{-i}) \geq u_i(\mathbf{y}^i, \mathbf{z}^{-i}), \tag{5.22}$$

and for at least one $\mathbf{z}^{-i}$, the inequality in Eq. (5.22) is strict.

Definition 5.31 (Strict Dominance). A mixed strategy, $\mathbf{x}^i \in \Delta_{n_i}$, for Player P_i *strictly dominates* another strategy, $\mathbf{y}^i \in \Delta_{n_i}$, for Player P_i if *for all* mixed strategies $\mathbf{z}^{-i}$, we have

$$u_i(\mathbf{x}^i, \mathbf{z}^{-i}) > u_i(\mathbf{y}^i, \mathbf{z}^{-i}). \tag{5.23}$$

Definition 5.32 (Dominated Strategy). A strategy, $\mathbf{y}^i \in \Delta_{n_i}$, for Player P_i is said to be *weakly (strictly) dominated* if there is a strategy, $\mathbf{x}^i \in \Delta_{n_i}$, that weakly (strictly) dominates it.

Remark 5.33. In a two-player matrix game $\mathcal{G} = (\mathbf{P}, \Sigma, \mathbf{A}, \mathbf{B})$, the mixed strategy $\mathbf{x} \in \Delta_m$ for Player 1 weakly dominates the strategy $\mathbf{y} \in \Delta_m$ if for all $\mathbf{z} \in \Delta_n$ (mixed strategies for Player 2) we have

$$\mathbf{x}^T \mathbf{A} \mathbf{z} \geq \mathbf{y}^T \mathbf{A} \mathbf{z} \tag{5.24}$$

and the inequality is strict for at least one $\mathbf{z} \in \Delta_n$. If $\mathbf{x}$ strictly dominates $\mathbf{y}$, then we have

$$\mathbf{x}^T \mathbf{A} \mathbf{z} > \mathbf{y}^T \mathbf{A} \mathbf{z}, \tag{5.25}$$

for all $\mathbf{z} \in \Delta_n$.

Example 5.34 (Prisoner's Dilemma). The following example is called *prisoner's dilemma* and is a classic example in game theory. There are several variations of this example, but they all have the same structure: Two criminals – we call them Bonnie and Clyde – commit a bank robbery. They hide the money and are driving around wondering what to do next when they are pulled over and arrested for carrying an illegal weapon. The police suspect Bonnie and Clyde of the bank robbery but do not have any hard evidence. They separate the prisoners and offer the following options to Bonnie:

(1) If neither Bonnie nor Clyde confess, they will each go to prison for 1 year for carrying the illegal weapon.
(2) If Bonnie confesses, but Clyde does not, then Bonnie can go free while Clyde will go to jail for 10 years.
(3) If Clyde confesses and Bonnie does not, then Bonnie will go to jail for 10 years while Clyde will go free.
(4) If both Bonnie and Clyde confess, then they will each go to jail for 5 years.

A similar offer is made to Clyde. The scenario is described by a two-player bimatrix game with the player set $\mathbf{P} = \{$Bonnie, Clyde$\}$, the strategy sets $\Sigma_1 = \Sigma_2 = \{$Don't Confess, Confess$\}$, and the payoff matrices

$$\mathbf{A} = \begin{bmatrix} -1 & -10 \\ 0 & -5 \end{bmatrix} \quad \text{and} \quad \mathbf{B} = \begin{bmatrix} -1 & 0 \\ -10 & -5 \end{bmatrix}.$$

Here, payoffs are given in negative years (for years lost to prison). Bonnie's matrix is $\mathbf{A}$, and Clyde's matrix is $\mathbf{B}$. The rows (columns) correspond to the strategies "Don't Confess" and "Confess." Thus, we see that if Bonnie does not confess and Clyde does (row 1, column 2), then Bonnie loses 10 years and Clyde loses 0 years.

We can show that the strategy Confess dominates Don't Confess for Bonnie. Recall that pure strategies correspond to standard basis vectors. We claim that $\mathbf{e}_2$ strictly dominates $\mathbf{e}_1$ for Bonnie. From Remark 5.33, we must show that

$$\mathbf{e}_2^T \mathbf{A} \mathbf{z} > \mathbf{e}_1^T \mathbf{A} \mathbf{z} \tag{5.26}$$

for all $\mathbf{z} \in \Delta_2$. We know that

$$\mathbf{z} = \begin{bmatrix} z_1 \\ z_2 \end{bmatrix},$$

where $z_1 + z_2 = 1$ and $z_1, z_2 \geq 0$. For simplicity, let's redefine

$$\mathbf{z} = \begin{bmatrix} z \\ (1-z) \end{bmatrix},$$

with $z \geq 0$. We know that

$$\mathbf{e}_2^T \mathbf{A} = \begin{bmatrix} 0 & 1 \end{bmatrix} \begin{bmatrix} -1 & -10 \\ 0 & -5 \end{bmatrix} = \begin{bmatrix} 0 & -5 \end{bmatrix},$$

$$\mathbf{e}_1^T \mathbf{A} = \begin{bmatrix} 1 & 0 \end{bmatrix} \begin{bmatrix} -1 & -10 \\ 0 & -5 \end{bmatrix} = \begin{bmatrix} -1 & -10 \end{bmatrix}.$$

Then,

$$\mathbf{e}_2^T \mathbf{A}\mathbf{z} = \begin{bmatrix} 0 & -5 \end{bmatrix} \begin{bmatrix} z \\ (1-z) \end{bmatrix} = -5(1-z) = 5z - 5,$$

$$\mathbf{e}_1^T \mathbf{A}\mathbf{z} = \begin{bmatrix} -1 & -10 \end{bmatrix} \begin{bmatrix} z \\ (1-z) \end{bmatrix} = -z - 10(1-z) = 9z - 10.$$

There are many ways to show that if $z \in [0,1]$, $5z - 5 > 9z - 10$, but the easiest way is to plot the two functions. This is shown in Fig. 5.5. It is also possible to show this analytically, but the visual proof is far more appealing. We can see the line corresponding to the payoff for Confess is always above the line corresponding to the payoff for Don't Confess.

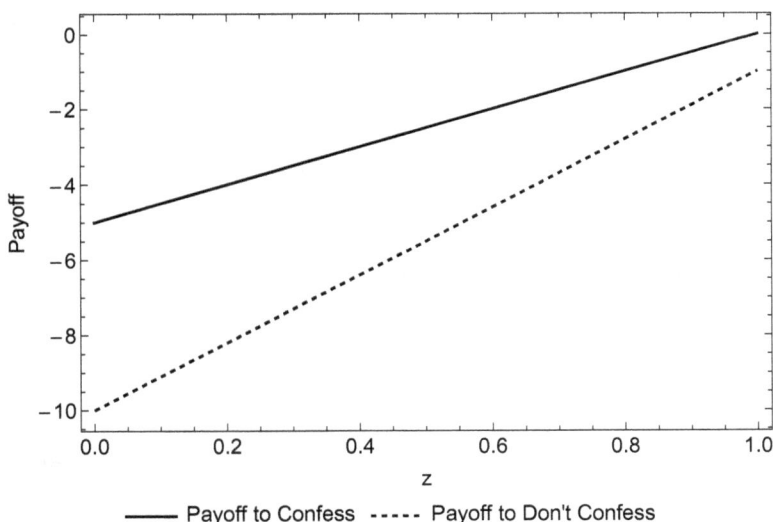

Fig. 5.5. Payoff for Bonnie's pure strategies for varying probabilities that Clyde confesses.

Remark 5.35. In general, prisoner's dilemma is presented as an abstract symmetric game, with Player 1's payoff matrix being

$$\mathbf{A} = \begin{bmatrix} R & S \\ T & P \end{bmatrix}$$

and Player 2's payoff matrix being $\mathbf{B} = \mathbf{A}^T$. We assume that $T > R > P > S$. In this way, row 2 of $\mathbf{A}$ strictly dominates row 1, while column 2 of $\mathbf{B}$ strictly dominates column 1.

Remark 5.36. Strict dominance can be extremely useful for identifying pure Nash equilibria. This is especially true for matrix games. We summarize this in the following two theorems.

Theorem 5.37. *Let* $\mathcal{G} = (\mathbf{P}, \Sigma, \mathbf{A}, \mathbf{B})$ *be a two-player matrix game, with* $\mathbf{A}, \mathbf{B} \in \mathbb{R}^{m \times n}$. *If*

$$\mathbf{e}_i^T \mathbf{A} \mathbf{e}_k > \mathbf{e}_j^T \mathbf{A} \mathbf{e}_k, \tag{5.27}$$

for $k \in \{1, \ldots, n\}$, *then* $\mathbf{e}_i$ *strictly dominates* $\mathbf{e}_j$ *for Player 1.*

Remark 5.38. We know that $\mathbf{e}_i^T \mathbf{A}$ is the ith row of $\mathbf{A}$. Theorem 5.37 states that if every element in $\mathbf{A}_i.$ (the ith row of $\mathbf{A}$) is greater than its corresponding element in $\mathbf{A}_j.$ (the jth row of $\mathbf{A}$), then Player 1's ith strategy strictly dominates Player 1's jth strategy.

Proof.   For all $k \in \{1, \ldots, n\}$, we know that

$$\mathbf{e}_i^T \mathbf{A} \mathbf{e}_k > \mathbf{e}_j^T \mathbf{A} \mathbf{e}_k.$$

Suppose that $z_1, \ldots, z_n \in [0, 1]$, with $z_1 + \cdots + z_n = 1$. Then, for each $k \in \{1, \ldots, n\}$, we know that

$$\mathbf{e}_i^T \mathbf{A} \mathbf{e}_k z_k > \mathbf{e}_j^T \mathbf{A} \mathbf{e}_k z_k.$$

Adding these inequalities together gives

$$\mathbf{e}_i^T \mathbf{A} \mathbf{e}_1 z_1 + \cdots + \mathbf{e}_i^T \mathbf{A} \mathbf{e}_n z_n > \mathbf{e}_j^T \mathbf{A} \mathbf{e}_1 z_1 + \cdots + \mathbf{e}_j^T \mathbf{A} \mathbf{e}_n z_n.$$

Factoring, we have

$$\mathbf{e}_i^T \mathbf{A} \left(z_1 \mathbf{e}_1 + \cdots + z_n \mathbf{e}_n \right) > \mathbf{e}_j^T \mathbf{A} \left(z_1 \mathbf{e}_1 + \cdots + z_n \mathbf{e}_n \right).$$

Define

$$\mathbf{z} = z_1 \mathbf{e}_1 + \cdots + z_n \mathbf{e}_n = \begin{bmatrix} z_1 \\ \vdots \\ z_n \end{bmatrix}.$$

Since the original $z_1, \ldots, z_n$ were chosen arbitrarily from $[0, 1]$ so that $z_1 + \cdots z_n = 1$, we know that

$$\mathbf{e}_i^T \mathbf{A} \mathbf{z} > \mathbf{e}_j^T \mathbf{A} \mathbf{z},$$

for all $\mathbf{z} \in \Delta_n$. Thus, $\mathbf{e}_i$ strictly dominates $\mathbf{e}_j$ by Definition 5.31. ☐

Remark 5.39. There is an analogous theorem for Player 2 which states that if each element of column $\mathbf{B}_{\cdot i}$ is greater than the corresponding element in column $\mathbf{B}_{\cdot j}$, then $\mathbf{e}_i$ strictly dominates strategy $\mathbf{e}_j$ for Player 2.

Remark 5.40. Theorem 5.37 can be generalized to N players; however, stating this theorem is notationally cumbersome and leads to no further insight.

Theorem 5.41. *Let $\mathcal{G} = (\mathbf{P}, \Sigma, \mathbf{A}, \mathbf{B})$ be a two-player matrix game. Suppose that the pure strategy $\mathbf{e}_j \in \Delta_m$ for Player 1 is strictly dominated by the pure strategy $\mathbf{e}_i \in \Delta_m$. If $(\mathbf{x}^*, \mathbf{y}^*)$ is a Nash equilibrium, then $x_j^* = 0$. Similarly, if the pure strategy $\mathbf{e}_j \in \Delta_n$ for Player 2 is strictly dominated by the pure strategy $\mathbf{e}_i \in \Delta_n$, then $y_j^* = 0$.*

Remark 5.42. Theorem 5.41 states that pure strategies that are dominated have no support (i.e., occur with a probability of zero) in a Nash equilibrium. We can use this fact in a process called analysis by iterated dominance to find pure strategy Nash equilibria and occasionally to simplify games with many strategies.

Proof of Theorem 5.41. We prove the theorem for Player 1; the proof for Player 2 is completely analogous. We proceed by contradiction. Let $\mathbf{x}^* = \langle x_1^*, \ldots, x_m^* \rangle$, and suppose that $x_j^* > 0$. We know that

$$\mathbf{e}_i^T \mathbf{A} \mathbf{y}^* > \mathbf{e}_j^* \mathbf{A} \mathbf{y}^*$$

because $\mathbf{e}_i$ strictly dominates $\mathbf{e}_j$. We can write

$$\mathbf{x}^{*T}\mathbf{A}\mathbf{y}^* = \left(x_1^*\mathbf{e}_1^T + \cdots + x_i^*\mathbf{e}_i^T + \cdots + x_j^*\mathbf{e}_j^T + \cdots + x_m^*\mathbf{e}_m^T\right)\mathbf{A}\mathbf{y}^*.$$
$$(5.28)$$

Since $x_j^* > 0$, we know that

$$x_j^*\mathbf{e}_i^T\mathbf{A}\mathbf{y}^* > x_j^*\mathbf{e}_j^*\mathbf{A}\mathbf{y}^*.$$

Replacing $x_j^*\mathbf{e}_j$ by $x_j^*\mathbf{e}_i$ in Eq. (5.28) yields the inequality

$$\left(x_1^*\mathbf{e}_1^T + \cdots + x_i^*\mathbf{e}_i^T + \cdots + x_j^*\mathbf{e}_i^T + \cdots + x_m^*\mathbf{e}_m^T\right)\mathbf{A}\mathbf{y}^*$$
$$> \left(x_1^*\mathbf{e}_1^T + \cdots + x_i^*\mathbf{e}_i^T + \cdots + x_j^*\mathbf{e}_j^T + \cdots + x_m^*\mathbf{e}_m^T\right)\mathbf{A}\mathbf{y}^*.$$
$$(5.29)$$

If we define $\mathbf{z} = \langle z_1, \ldots, z_m \rangle \in \Delta_m$ so that

$$z_k = \begin{cases} x_i^* + x_j^* & k = i \\ 0 & k = j \\ x_k & \text{otherwise,} \end{cases} \qquad (5.30)$$

then Eq. (5.29) implies

$$\mathbf{z}^T\mathbf{A}\mathbf{y}^* > \mathbf{x}^{*T}\mathbf{A}\mathbf{y}^*. \qquad (5.31)$$

Thus, $(\mathbf{x}^*, \mathbf{y}^*)$ could not have been a Nash equilibrium. This completes the proof. □

Remark 5.43. The preceding proof worked by showing that transferring the probability placed on strategy $\mathbf{e}_j$ to strategy $\mathbf{e}_i$ improved the payoff to Player 1 precisely because we assumed that $\mathbf{e}_i$ dominates $\mathbf{e}_j$. Consequently, no Nash equilibrium strategy can assign non-zero probability to a dominated strategy.

Example 5.44. We can use the two previous theorems to our advantage. Consider the prisoner's dilemma (Example 5.34). The payoff matrices (again) are

$$\mathbf{A} = \begin{bmatrix} -1 & -10 \\ 0 & -5 \end{bmatrix} \quad \text{and} \quad \mathbf{B} = \begin{bmatrix} -1 & 0 \\ -10 & -5 \end{bmatrix}.$$

For Bonnie, row (strategy) 1 is strictly dominated by row (strategy) 2. Thus, Bonnie will never play strategy 1 (Don't Confess) in a

Nash equilibrium. That is,

$$\mathbf{A}_{1\cdot} < \mathbf{A}_{2\cdot} \equiv \begin{bmatrix} -1 & -10 \end{bmatrix} < \begin{bmatrix} 0 & -5 \end{bmatrix}.$$

We can consider a new game in which we remove this strategy for Bonnie (since Bonnie will never play this strategy). The new game has $\mathbf{P} = \{\text{Bonnie}, \text{Clyde}\}$, $\Sigma_1 = \{\text{Confess}\}$, $\Sigma_2 = \{\text{Don't Confess}, \text{Confess}\}$. The new game matrices are

$$\mathbf{A}' = \begin{bmatrix} 0 & -5 \end{bmatrix} \quad \text{and} \quad \mathbf{B}' = \begin{bmatrix} -10 & -5 \end{bmatrix}.$$

In this new game, we note that for Clyde (Player 2), column (strategy) 2 strictly dominates column (strategy) 1. That is,

$$\mathbf{B}'_{\cdot 1} < \mathbf{B}'_{\cdot 2} \equiv -10 < -5.$$

Clyde will never play Strategy 1 (Don't Confess) in a Nash equilibrium. We can construct a new game with $\mathbf{P} = \{\text{Bonnie}, \text{Clyde}\}$, $\Sigma_1 = \{\text{Confess}\}$, $\Sigma_2 = \{\text{Confess}\}$ and (trivial) payoff matrices as

$$\mathbf{A}'' = -5 \quad \text{and} \quad \mathbf{B}'' = -5.$$

In this game, there is only one Nash equilibrium in which both players confess, and this equilibrium is the Nash equilibrium of the original game.

Remark 5.45 (Iterative Dominance). A game whose Nash equilibrium is computed using the method from Example 5.44, in which strictly dominated strategies are iteratively eliminated for the two players, is said to be *solved by iterated dominance*. A game that can be analyzed in this way is said to be *strictly dominance solvable*.

5.5   The Indifference Theorem

Theorem 5.46 (Indifference Theorem). *Let* $\mathcal{G} = (\mathbf{P}, \Sigma, \mathbf{A}, \mathbf{B})$ *be a two-player matrix game, and suppose that* $(\mathbf{x}^*, \mathbf{y}^*)$ *is a Nash equilibrium. If* $x_i^* > 0$, *then*

$$\mathbf{x}^{*T}\mathbf{A}\mathbf{y}^* = \mathbf{e}_i^T\mathbf{A}\mathbf{y}^*.$$

Likewise, if $y_j^* > 0$, *then*

$$\mathbf{x}^{*T}\mathbf{A}\mathbf{y}^* = \mathbf{x}^{*T}\mathbf{A}\mathbf{e}_j.$$

Remark 5.47. The indifference theorem states that if Player 2 plays a Nash equilibrium, then Player 1 is indifferent between playing her mixed-strategy Nash equilibrium and some pure strategy that has a non-zero probability in the Nash equilibrium. That is, Player 1 will receive the same payoff no matter whether she plays the mixed-strategy equilibrium or a pure-strategy equilibrium. This does not mean that both players can switch to a pure strategy equilibrium, as that could clearly change the payoffs.

Remark 5.48. The proof uses the same trick we have already used twice, though it does require a bit more intricacy because we will be ignoring those strategies that have zero probability in the Nash equilibrium.

Proof of the Indifference Theorem. We prove this for Player 1. The result follows by symmetry for Player 2. We know that we cannot have

$$\mathbf{e}_i^T \mathbf{A} \mathbf{y}^* > \mathbf{x}^{*T} \mathbf{A} \mathbf{y}^*$$

by the definition of a Nash equilibrium. Let $\mathcal{I} \subseteq \{1, \ldots, m\}$ be the set of indices so that $k \in \mathcal{I}$ if and only if $x_k^* > 0$. That is, $\mathcal{I}$ is the set of strategy indexes that occur with non-zero probability. We know that

$$\sum_{k \in \mathcal{I}} x_k^* = 1.$$

By construction,

$$\sum_{k \in \mathcal{I}} x_k^* \mathbf{e}_i = \sum_i x_k \mathbf{e}_i = \mathbf{x}^*,$$

which is simply a variation of Eq. (5.28) written more compactly. Assume that

$$\mathbf{e}_i^T \mathbf{A} \mathbf{y}^* < \mathbf{x}^{*T} \mathbf{A} \mathbf{y}^*, \tag{5.32}$$

for some $i \in \mathcal{I}$. We also know that

$$\mathbf{e}_k^T \mathbf{A} \mathbf{y}^* \leq \mathbf{x}^{*T} \mathbf{A} \mathbf{y}^*,$$

for all $k \in \mathcal{I}$, by the definition of the Nash equilibrium. Since $x_k^* > 0$ for all $k \in \mathcal{I}$, we can rewrite these inequalities as

$$x_i^* \mathbf{e}_i^T \mathbf{A} \mathbf{y}^* < x_i^* \mathbf{x}^{*T} \mathbf{A} \mathbf{y}^*,$$

$$x_k^* \mathbf{e}_k^T \mathbf{A} \mathbf{y}^* \leq x_k^* \mathbf{x}^{*T} \mathbf{A} \mathbf{y}^* \quad \forall k \in \mathcal{I}, k \neq i.$$

Adding all the inequalities yields

$$x_i^* \mathbf{e}_i^T \mathbf{A} \mathbf{y}^* + \sum_{j \in \mathcal{I}, k \neq i} x_k^* \mathbf{e}_k^T \mathbf{A} \mathbf{y}^* = \left(\sum_{k \in \mathcal{I}} x_k^* \mathbf{e}_k \right) \mathbf{A} \mathbf{y}^* = \mathbf{x}^{*T} \mathbf{A} \mathbf{y}^*$$

$$< x_i \mathbf{x}^{*T} \mathbf{A} \mathbf{y}^* + \sum_{k \in \mathcal{I}, k \neq i} x_k \mathbf{x}^{*T} \mathbf{A} \mathbf{y}^* = \left(\sum_{k \in \mathcal{I}} x_k^* \right) \mathbf{x}^{*T} \mathbf{A} \mathbf{y}^* = \mathbf{x}^{*T} \mathbf{A} \mathbf{y}^*.$$

The strictness of the inequality follows from our assumption in Eq. (5.32). But this is a contradiction. Therefore, we know that, for all $i \in \mathcal{I}$, we cannot have $\mathbf{e}_i^T \mathbf{A} \mathbf{y}^* < \mathbf{x}^{*T} \mathbf{A} \mathbf{y}^*$; therefore, it follows that we must have $\mathbf{e}_i^T \mathbf{A} \mathbf{y}^* = \mathbf{x}^{*T} \mathbf{A} \mathbf{y}^*$ for all $i \in \mathcal{I}$. The argument for Player 2 is identical. This completes the proof. $\square$

Remark 5.49. Note that the strategies with zero probability are free to yield a lower payoff without affecting the proof. That is, if $k \neq \mathcal{I}$ in the proof above, it is perfectly fine that

$$\mathbf{e}_k^T \mathbf{A} \mathbf{y}^* < \mathbf{x}^{*T} \mathbf{A} \mathbf{y}^*$$

because $x_k^* = 0$, and so those terms will never appear in any of the sums used in the proof.

Example 5.50. Recall that the payoff matrix for the zero-sum game rock-paper-scissors is

$$\mathbf{A} = \begin{bmatrix} 0 & -1 & 1 \\ 1 & 0 & -1 \\ -1 & 1 & 0 \end{bmatrix}.$$

The Nash equilibrium is $\mathbf{x}^* = \mathbf{y}^* = \langle \frac{1}{3}, \frac{1}{3}, \frac{1}{3} \rangle$. Note that

$$\mathbf{A} \mathbf{y}^* = \begin{bmatrix} 0 & -1 & 1 \\ 1 & 0 & -1 \\ -1 & 1 & 0 \end{bmatrix} \begin{bmatrix} \frac{1}{3} \\ \frac{1}{3} \\ \frac{1}{3} \end{bmatrix} = \begin{bmatrix} 0 \\ 0 \\ 0 \end{bmatrix}.$$

Thus, it is easy to see that the indifference theorem holds in this case. In fact, for any $\mathbf{x} \in \Delta_3$, we see at once that $\mathbf{x}\mathbf{A}\mathbf{y}^* = 0$.

Remark 5.51. In light of the previous example, we see that the indifference theorem can be generalized a bit. This generalization appears in the exercises.

5.6   The Minimax Theorem

Remark 5.52. We now return to zero-sum games and show that there is a Nash equilibrium for every zero-sum game. Before proceeding, we recall the definition of a Nash equilibrium as it applies to a zero-sum game. A mixed strategy, $(\mathbf{x}^*, \mathbf{y}^*) \in \Delta$, is a Nash equilibrium for a zero-sum game, $\mathcal{G} = (\mathbf{P}, \Sigma, \mathbf{A})$ with $\mathbf{A} \in \mathbb{R}^{m \times n}$, if we have

$$\mathbf{x}^{*T} \mathbf{A} \mathbf{y}^* \geq \mathbf{x}^T \mathbf{A} \mathbf{y}^*,$$

for all $\mathbf{x} \in \Delta_m$ and

$$\mathbf{x}^{*T} \mathbf{A} \mathbf{y}^* \leq \mathbf{x}^{*T} \mathbf{A} \mathbf{y},$$

for all $\mathbf{y} \in \Delta_n$.

Remark 5.53. Let $\mathcal{G} = (\mathbf{P}, \Sigma, \mathbf{A})$ be a zero-sum game, with $\mathbf{A} \in \mathbb{R}^{m \times n}$. Define the function $v_1 \colon \Delta_m \to \mathbb{R}$ as

$$v_1(\mathbf{x}) = \min_{\mathbf{y} \in \Delta_n} \mathbf{x}^T \mathbf{A} \mathbf{y}. \tag{5.33}$$

That is, given $\mathbf{x} \in \Delta_m$, we choose a vector $\mathbf{y}$ that minimizes the value $\mathbf{x}^T \mathbf{A} \mathbf{y}$. This value is the best possible result Player 1 can expect if she announces to Player 2 that she will play strategy $\mathbf{x}$. Player 1 then faces the problem that she would like to maximize this value by choosing $\mathbf{x}$ appropriately. That is, Player 1 hopes to solve the problem

$$\max_{\mathbf{x} \in \Delta_m} v_1(\mathbf{x}). \tag{5.34}$$

Thus, Player 1's problem is to solve

$$\max_{\mathbf{x} \in \Delta_m} v_1(\mathbf{x}) = \max_{\mathbf{x}} \min_{\mathbf{y}} \mathbf{x}^T \mathbf{A} \mathbf{y}. \tag{5.35}$$

By a similar argument, define the function $v_2 \colon \Delta_n \to \mathbb{R}$ as

$$v_2(\mathbf{y}) = \max_{\mathbf{x} \in \Delta_m} \mathbf{x}^T \mathbf{A} \mathbf{y}. \tag{5.36}$$

That is, given $\mathbf{y} \in \Delta_n$, we choose a vector $\mathbf{x}$ that maximizes $\mathbf{x}^T \mathbf{A} \mathbf{y}$. This value is the best possible result that Player 2 can expect if he announces to Player 1 that he will play strategy $\mathbf{y}$. Player 2 then faces the problem that he would like to minimize this value by choosing $\mathbf{y}$ appropriately. That is, Player 2 hopes to solve the problem

$$\min_{\mathbf{y} \in \Delta_n} v_2(\mathbf{y}). \tag{5.37}$$

Player 2's problem is to solve

$$\min_{\mathbf{y} \in \Delta_n} v_2(\mathbf{y}) = \min_{\mathbf{y}} \max_{\mathbf{x}} \mathbf{x}^T \mathbf{A} \mathbf{y}. \tag{5.38}$$

Note that this is the precise analogue in mixed strategies to the concept of a saddle point. The functions v_1 and v_2 are called the value functions for Players 1 and 2, respectively. The main problem we must tackle now is to determine whether these maximization and minimization problems can be solved.

Remark 5.54. The proof of the following lemma is left as an exercise.

Lemma 5.55. *Let* $\mathcal{G} = (\mathbf{P}, \Sigma, \mathbf{A})$ *be a zero-sum game, with* $\mathbf{A} \in \mathbb{R}^{m \times n}$. *Then,*

$$\max_{\mathbf{x} \in \Delta_m} v_1(\mathbf{x}) \leq \min_{\mathbf{y} \in \Delta_n} v_2(\mathbf{y}). \tag{5.39}$$

Remark 5.56. The proof of the following theorem, the minimax theorem, is long and uses some trickery. It is best to read it a few times, or to skip it on the first reading, since the insight is all in the theorem rather than in the proof.

Theorem 5.57 (Minimax Theorem). *Let* $\mathcal{G} = (\mathbf{P}, \Sigma, \mathbf{A})$ *be a zero-sum game, with* $\mathbf{A} \in \mathbb{R}^{m \times n}$. *Then, the following are equivalent:*

(1) *There is a Nash equilibrium* $(\mathbf{x}^*, \mathbf{y}^*)$ *for* $\mathcal{G}$.
(2) *The following equation holds:*

$$v_1 = \max_{\mathbf{x}} \min_{\mathbf{y}} \mathbf{x}^T \mathbf{A} \mathbf{y} = \min_{\mathbf{y}} \max_{\mathbf{x}} \mathbf{x}^T \mathbf{A} \mathbf{y} = v_2. \tag{5.40}$$

(3) *There exists a real number* v *and* $\mathbf{x}^* \in \Delta_m$ *and* $\mathbf{y}^* \in \Delta_n$ *so that the following inequalities hold:*

$$\sum_i \mathbf{A}_{ij} \mathbf{x}_i^* \geq v, \quad \text{with } j \in \{1, \dots, n\} \quad \text{and}$$

$$\sum_j \mathbf{A}_{ij} \mathbf{y}_j^* \leq v, \quad \text{with } i \in 1, \dots, m\}.$$

Proof.   (A version of this proof is given in Ref. [1], Appendix 2.)

($1 \implies 2$): Suppose that $(\mathbf{x}^*, \mathbf{y}^*) \in \Delta$ is a Nash equilibrium. By the definition of a minimum, we know that

$$v_2 = \min_{\mathbf{y}} \max_{\mathbf{x}} \mathbf{x}^T \mathbf{A} \mathbf{y} \leq \max_{\mathbf{x}} \mathbf{x}^T \mathbf{A} \mathbf{y}^*.$$

The fact that for all $\mathbf{x} \in \Delta_m$,

$$\mathbf{x}^{*T} \mathbf{A} \mathbf{y}^* \geq \mathbf{x}^T \mathbf{A} \mathbf{y}^*,$$

implies that

$$\mathbf{x}^{*T} \mathbf{A} \mathbf{y}^* = \max_{\mathbf{x}} \mathbf{x}^T \mathbf{A} \mathbf{y}^*.$$

Thus, we have

$$v_2 = \min_{\mathbf{y}} \max_{\mathbf{x}} \mathbf{x}^T \mathbf{A} \mathbf{y} \leq \max_{\mathbf{x}} \mathbf{x}^T \mathbf{A} \mathbf{y}^* = \mathbf{x}^{*T} \mathbf{A} \mathbf{y}^*.$$

Again, the fact that for all $\mathbf{y} \in \Delta_n$,

$$\mathbf{x}^{*T} \mathbf{A} \mathbf{y}^* \leq \mathbf{x}^{*T} \mathbf{A} \mathbf{y},$$

implies that

$$\mathbf{x}^{*T} \mathbf{A} \mathbf{y}^* = \min_{\mathbf{y}} \mathbf{x}^{*T} \mathbf{A} \mathbf{y}.$$

Thus,

$$v_2 = \min_{\mathbf{y}} \max_{\mathbf{x}} \mathbf{x}^T \mathbf{A} \mathbf{y} \leq \max_{\mathbf{x}} \mathbf{x}^T \mathbf{A} \mathbf{y}^* = \mathbf{x}^{*T} \mathbf{A} \mathbf{y}^* = \min_{\mathbf{y}} \mathbf{x}^{*T} \mathbf{A} \mathbf{y}.$$

Finally, by the definition of a maximum, we know that

$$v_2 = \min_{\mathbf{y}} \max_{\mathbf{x}} \mathbf{x}^T \mathbf{A} \mathbf{y} \leq \max_{\mathbf{x}} \mathbf{x}^T \mathbf{A} \mathbf{y}^* = \mathbf{x}^{*T} \mathbf{A} \mathbf{y}^*$$

$$= \min_{\mathbf{y}} \mathbf{x}^{*T} \mathbf{A} \mathbf{y} \leq \max_{\mathbf{x}} \min_{\mathbf{y}} \mathbf{x}^T \mathbf{A} \mathbf{y} = v_1. \tag{5.41}$$

By Lemma 5.55, we know that $v_1 \leq v_2$. We have just proved that $v_2 \leq v_1$. Therefore, $v_1 = v_2$, as required.

($2 \implies 3$): Let $v = v_1 = v_2$, let $\mathbf{x}^*$ be the vector that maximizes $v_1(\mathbf{x})$, and let $\mathbf{y}^*$ be the vector that minimizes $v_2(\mathbf{y})$. For fixed j, we know that

$$\sum_i \mathbf{A}_{ij} \mathbf{x}_i^* = \mathbf{x}^{*T} \mathbf{A} \mathbf{e}_j.$$

Note that we are summing down a column of $\mathbf{A}$ in the preceding equation. By the definition of minimum, we know that

$$\sum_i \mathbf{A}_{ij}\mathbf{x}_i^* = \mathbf{x}^{*T}\mathbf{A}\mathbf{e}_j \geq \min_{\mathbf{y}} \mathbf{x}^{*T}\mathbf{A}\mathbf{y}.$$

We defined $\mathbf{x}^*$ so that it is the maximin value, and thus,

$$\sum_i \mathbf{A}_{ij}\mathbf{x}_i^* = \mathbf{x}^{*T}\mathbf{A}\mathbf{e}_j \geq \min_{\mathbf{y}} \mathbf{x}^{*T}\mathbf{A}\mathbf{y} = \max_{\mathbf{x}} \min_{\mathbf{y}} \mathbf{x}^T\mathbf{A}\mathbf{y}$$

$$= v = \min_{\mathbf{y}} \max_{\mathbf{x}} \mathbf{x}^T\mathbf{A}\mathbf{y}.$$

By a similar argument, we defined $\mathbf{y}^*$ so that it is the minimax value, and thus,

$$\sum_i \mathbf{A}_{ij}\mathbf{x}_i^* = \mathbf{x}^{*T}\mathbf{A}\mathbf{e}_j \geq \min_{\mathbf{y}} \mathbf{x}^{*T}\mathbf{A}\mathbf{y} = \max_{\mathbf{x}} \min_{\mathbf{y}} \mathbf{x}^T\mathbf{A}\mathbf{y}$$

$$= v = \min_{\mathbf{y}} \max_{\mathbf{x}} \mathbf{x}^T\mathbf{A}\mathbf{y} = \max_{\mathbf{x}} \mathbf{x}^T\mathbf{A}\mathbf{y}^*.$$

Finally, for fixed i, we know that

$$\sum_j \mathbf{A}_{ij}\mathbf{y}_j^* = \mathbf{e}_i^T\mathbf{A}\mathbf{y}^*,$$

and thus, we conclude that

$$\sum_i \mathbf{A}_{ij}\mathbf{x}_i^* = \mathbf{x}^{*T}\mathbf{A}\mathbf{e}_j \geq \min_{\mathbf{y}} \mathbf{x}^{*T}\mathbf{A}\mathbf{y} = \max_{\mathbf{x}} \min_{\mathbf{y}} \mathbf{x}^T\mathbf{A}\mathbf{y}$$

$$= v = \min_{\mathbf{y}} \max_{\mathbf{x}} \mathbf{x}^T\mathbf{A}\mathbf{y} = \max_{\mathbf{x}} \mathbf{x}^T\mathbf{A}\mathbf{y}^* \geq \mathbf{e}_i^T\mathbf{A}\mathbf{y}^* = \sum_j \mathbf{A}_{ij}\mathbf{y}_j^*.$$

$$(5.42)$$

From Eq. (5.42), we can read off the two inequalities

$$\sum_i \mathbf{A}_{ij}\mathbf{x}_i^* \geq v \quad \text{for } j \in \{1,\ldots,n\},$$

$$\sum_j \mathbf{A}_{ij}\mathbf{y}_j^* \leq v \quad \text{for } i \in \{1,\ldots,m\}.$$

$(3 \implies 1)$: For any fixed j, we know that

$$\mathbf{x}^{*T}\mathbf{A}\mathbf{e}_j \geq v.$$

Thus, if $y_1, \ldots, y_n \in [0,1]$ and $y_1 + \cdots + y_n = 1$ for each $j \in \{1, \ldots, n\}$, we know that

$$\mathbf{x}^{*T}\mathbf{A}\mathbf{e}_j y_j \geq v y_j.$$

Thus, we can conclude that

$$\mathbf{x}^{*T}\mathbf{A}\mathbf{e}_1 y_1 + \cdots + \mathbf{x}^{*T}\mathbf{A}\mathbf{e}_n y_n = \mathbf{x}^{*T}\mathbf{A}\left(\mathbf{e}_1 y_1 + \cdots + \mathbf{e}_n y_n\right) \geq v.$$

Letting $\mathbf{y} = \langle y_1, \ldots, y_n \rangle$, we can conclude that

$$\mathbf{x}^{*T}\mathbf{A}\mathbf{y} \geq v, \tag{5.43}$$

for any $\mathbf{y} \in \Delta_n$. By a similar argument, we know that

$$\mathbf{x}^T\mathbf{A}\mathbf{y}^* \leq v, \tag{5.44}$$

for all $\mathbf{x} \in \Delta_m$. From Eq. (5.44), we conclude that

$$\mathbf{x}^{*T}\mathbf{A}\mathbf{y}^* \leq v, \tag{5.45}$$

and from Eq. (5.43), we conclude that

$$\mathbf{x}^{*T}\mathbf{A}\mathbf{y}^* \geq v. \tag{5.46}$$

Thus, $v = \mathbf{x}^{*T}\mathbf{A}\mathbf{y}^*$, and we know that, for all $\mathbf{x}$ and $\mathbf{y}$,

$$\mathbf{x}^{*T}\mathbf{A}\mathbf{y}^* \geq \mathbf{x}^T\mathbf{A}\mathbf{y}^*,$$

$$\mathbf{x}^{*T}\mathbf{A}\mathbf{y}^* \leq \mathbf{x}^{*T}\mathbf{A}\mathbf{y}.$$

Thus, $(\mathbf{x}^*, \mathbf{y}^*)$ is a Nash equilibrium. This completes the proof. □

Remark 5.58. Theorem 5.57 does not assert the existence of a Nash equilibrium; it simply provides insight into what happens if one exists. In particular, we know that the game has a unique value:

$$v = \max_{\mathbf{x}} \min_{\mathbf{y}} \mathbf{x}^T\mathbf{A}\mathbf{y} = \min_{\mathbf{y}} \max_{\mathbf{x}} \mathbf{x}^T\mathbf{A}\mathbf{y}. \tag{5.47}$$

Proving the existence of a Nash equilibrium can be accomplished in several ways, the oldest of which uses a topological argument, which we present in the following. We can also use a linear programming-based argument, which we explore in Chapter 7.

5.7 Existence of Nash Equilibria

Lemma 5.59 (Brouwer Fixed Point Theorem). *Let Δ be the mixed strategy space of a two-player zero-sum game. If $T\colon \Delta \to \Delta$ is continuous, then there exists a pair of strategies, $(\mathbf{x}^*, \mathbf{y}^*)$, so that $T(\mathbf{x}^*, \mathbf{y}^*) = (\mathbf{x}^*, \mathbf{y}^*)$. That is, $(\mathbf{x}^*, \mathbf{y}^*)$ is a **fixed point** of the mapping T.*

Remark 5.60. In the previous lemma, we are casually avoiding a formal definition of function continuity and relying on the fact that the reader has probably encountered this definition in a calculus class. Moreover, the proof of Brouwer's fixed point theorem is well outside the scope of this book. It is a deep theorem in topology. The interested reader should consult Ref. [61] (pp. 351–353), which also has a definition of function continuity.

Remark 5.61. Before proving that every zero-sum game has a Nash equilibrium, we state a lemma. The proof is left as an exercise.

Lemma 5.62. *Let $\mathcal{G} = (\mathbf{P}, \Sigma, \mathbf{A})$ be a zero-sum game, with $\mathbf{A} \in \mathbb{R}^{m \times n}$. Let $\mathbf{x}^* \in \Delta_m$ and $\mathbf{y}^* \in \Delta_n$. If*

$$\mathbf{x}^{*T} \mathbf{A} \mathbf{y}^* \geq \mathbf{e}_i^T \mathbf{A} \mathbf{y}^*,$$

for all $i \in \{1, \ldots, m\}$, and

$$\mathbf{x}^{*T} \mathbf{A} \mathbf{y}^* \leq \mathbf{x}^{*T} \mathbf{A} \mathbf{e}_j,$$

for all $j \in \{1, \ldots, n\}$, then $(\mathbf{x}^*, \mathbf{y}^*)$ is an equilibrium.

Theorem 5.63. *Let $\mathcal{G} = (\mathbf{P}, \Sigma, \mathbf{A})$ be a zero-sum game with $\mathbf{A} \in \mathbb{R}^{m \times n}$. Then, there is a Nash equilibrium, $(\mathbf{x}^*, \mathbf{y}^*)$.*

Nash's Proof. (A version of this proof is given in Ref. [1], Appendix 2.) Let $(\mathbf{x}, \mathbf{y}) \in \Delta$ be a pair of mixed strategies for Players 1 and 2. Define the following functions:

$$c_i(\mathbf{x}, \mathbf{y}) = \begin{cases} \mathbf{e}_i^T \mathbf{A} \mathbf{y} - \mathbf{x}^T \mathbf{A} \mathbf{y} & \text{if this quantity is positive} \\ 0 & \text{otherwise} \end{cases} \tag{5.48}$$

$$d_j(\mathbf{x}, \mathbf{y}) = \begin{cases} \mathbf{x}^T \mathbf{A} \mathbf{y} - \mathbf{x}^T \mathbf{A} \mathbf{e}_j & \text{if this quantity is positive} \\ 0 & \text{otherwise,} \end{cases} \tag{5.49}$$

for $i \in \{1, \ldots, m\}$ and $j \in \{1, \ldots, n\}$. Note that c_i (respectively, d_j) is non-zero just in case the pure strategy i (resp., j) offers a better payoff to Player 1 (resp., Player 2) than the strategy $\mathbf{x}$ (resp., $\mathbf{y}$).

Let $T \colon \Delta \to \Delta$, where $T(\mathbf{x}, \mathbf{y}) = (\mathbf{x}', \mathbf{y}')$, so that for $i \in \{1, \ldots, m\}$, we have

$$x_i' = \frac{x_i + c_i(\mathbf{x}, \mathbf{y})}{1 + \sum_{k=1}^m c_k(\mathbf{x}, \mathbf{y})}, \tag{5.50}$$

and for $j \in \{1, \ldots, n\}$, we have

$$y_j' = \frac{y_j + d_j(\mathbf{x}, \mathbf{y})}{1 + \sum_{k=1}^n d_k(\mathbf{x}, \mathbf{y})}. \tag{5.51}$$

Since $x_1 + x_2 + \cdots + x_m = 1$, we know that

$$x_1' + \cdots + x_m' = \frac{x_1 + \cdots + x_m + \sum_{k=1}^m c_k(\mathbf{x}, \mathbf{y})}{1 + \sum_{k=1}^m c_k(\mathbf{x}, \mathbf{y})} = 1. \tag{5.52}$$

It is also clear that since $x_i \geq 0$ for all $i \in \{1, \ldots, m\}$, we know that $x_i' \geq 0$ for all i. A similar argument shows that $y_j' \geq 0$ for all $j \in \{1, \ldots, n\}$ and $y_1' + y_2' + \cdots + y_n' = 1$. Thus, as we have defined it: T is a map from Δ to Δ. The fact that T is continuous follows from the continuity of the payoff function. Now, we show that $(\mathbf{x}, \mathbf{y})$ is a Nash equilibrium if and only if it is a fixed point of T.

To see this, note that $c_i(\mathbf{x}, \mathbf{y})$ measures the amount that the pure strategy $\mathbf{e}_i$ is better than $\mathbf{x}$ as a response to $\mathbf{y}$. That is, if Player 2 decides to play strategy $\mathbf{y}$, then $c_i(\mathbf{x}, \mathbf{y})$ tells us if and how much playing the pure strategy $\mathbf{e}_i$ is better than playing $\mathbf{x} \in \Delta_m$. Similarly, $d_j(\mathbf{x}, \mathbf{y})$ measures how much better $\mathbf{e}_j$ is as a response to Player 1's strategy $\mathbf{x}$ than the strategy $\mathbf{y}$ for Player 2. Suppose that $(\mathbf{x}^*, \mathbf{y}^*)$ is a Nash equilibrium. Then, necessarily, $c_i(\mathbf{x}^*, \mathbf{y}^*) = 0 = d_j(\mathbf{x}^*, \mathbf{y}^*)$ for all i and j, by the definition of equilibrium. Thus, $x_i' = x_i^*$ for all i and $y_j' = y_j^*$ for all j. Thus, we have shown that $(\mathbf{x}^*, \mathbf{y}^*)$ is a fixed point of T.

To show the converse, suppose that $(\mathbf{x}, \mathbf{y})$ is a fixed point of T. It suffices to show that there is at least one i so that $x_i > 0$ and $c_i(\mathbf{x}, \mathbf{y}) = 0$. We know that there is at least one i for which $x_i > 0$

because $x_1 + \cdots + x_m = 1$. Note that

$$\mathbf{x}^T \mathbf{A} \mathbf{y} = \sum_{i=1}^{m} x_i \mathbf{e}_i^T \mathbf{A} \mathbf{y}.$$

Thus, $\mathbf{x}^T \mathbf{A} \mathbf{y} < \mathbf{e}_i^T \mathbf{A} \mathbf{y}$ cannot hold for all $i \in \{1, \ldots, m\}$ with $x_i > 0$ (otherwise, the previous equation would not hold). Therefore, for at least one i with $x_i > 0$, we must have $c_i(\mathbf{x}, \mathbf{y}) = 0$. But for this specific i, the fact that $(\mathbf{x}, \mathbf{y})$ is a fixed point implies that

$$x_i = \frac{x_i}{1 + \sum_{k=1}^{m} c_k(\mathbf{x}, \mathbf{y})}. \tag{5.53}$$

This implies that

$$\sum_{k=1}^{m} c_k(\mathbf{x}, \mathbf{y}) = 0.$$

For this to be true, we conclude that $c_k(\mathbf{x}, \mathbf{y}) = 0$ because $c_k(\mathbf{x}, \mathbf{y}) \geq 0$ for all $k \in \{1, \ldots, m\}$. A similar argument applies to $\mathbf{y}$. Thus, we know that $c_i(\mathbf{x}, \mathbf{y}) = 0 = d_j(\mathbf{x}, \mathbf{y})$ for all $i \in \{1, \ldots, m\}$ and $j \in \{1, \ldots, n\}$. We have shown that

$$\mathbf{x}^T \mathbf{A} \mathbf{y} \geq \mathbf{e}_i^T \mathbf{A} \mathbf{y} \quad \text{and} \quad \mathbf{x}^T \mathbf{A} \mathbf{y} \leq \mathbf{x}^T \mathbf{A} \mathbf{e}_j,$$

for all $i \in \{1, \ldots, m\}$ and $j \in \{1, \ldots, n\}$. Therefore, by Lemma 5.62, we have that $(\mathbf{x}, \mathbf{y})$ is a Nash equilibrium.

Now, apply Lemma 5.59 (Brouwer's fixed point theorem) to see that T must have a fixed point, and thus every two-player zero-sum game has a Nash equilibrium. This completes the proof. □

5.8   Finding Nash Equilibria in Simple Games

Remark 5.64. It is relatively straightforward to find a Nash equilibrium in 2×2 zero-sum games, assuming that a saddle point cannot be identified using the approach from Example 5.3 or the method of iterated dominance. We illustrate this approach using the *Battle of Avranches*, which we know does not have an equilibrium in pure strategies.

Example 5.65. Consider the Battle of Avranches (Example 5.11). The payoff matrix is

$$\mathbf{A} = \begin{bmatrix} 2 & 3 \\ 1 & 5 \\ 6 & 4 \end{bmatrix}.$$

Note first that Row 1 (Bradley' first strategy) is strictly dominated by Row 3 (Bradley's third strategy), and thus we can reduce the payoff matrix to

$$\mathbf{A} = \begin{bmatrix} 1 & 5 \\ 6 & 4 \end{bmatrix}.$$

Suppose that Bradley chooses the strategy

$$\mathbf{x} = \begin{bmatrix} x \\ 1 - x \end{bmatrix}$$

with $x \in [0, 1]$. If von Kluge chooses to attack (column one), then Bradley's expected payoff will be

$$\mathbf{x}^T \mathbf{A} \mathbf{e}_1 = \begin{bmatrix} x & 1 - x \end{bmatrix} \begin{bmatrix} 1 & 5 \\ 6 & 4 \end{bmatrix} \begin{bmatrix} 1 \\ 0 \end{bmatrix} = x + 6(1 - x) = 6 - 5x.$$

A similar argument shows that if von Kluge chooses to retreat (column two), then Bradley's expected payoff will be

$$\mathbf{x}^T \mathbf{A} \mathbf{e}_2 = 5x + 4(1 - x) = x + 4.$$

We can visualize these two possibilities by plotting the functions, as shown in Fig. 5.6 (left). Reasoning that von Kluge will attempt to minimize his payoff, Bradley should consider the function

$$u_1(x) = \min\{6 - 5x, x + 4\}$$

and find a value x that maximizes $u_1(x)$. This maximizing point comes at $x = \frac{1}{3}$, where the two lines intersect.

Put another way, when $x \le \frac{1}{3}$, von Kluge does better if he retreats because $x + 4$ is below $6 - 5x$ (remember, von Kluge wishes to minimize Bradley's payoff). That is, the best Bradley can hope to get is $x + 4$ if he announces to von Kluge that he is playing a mixed

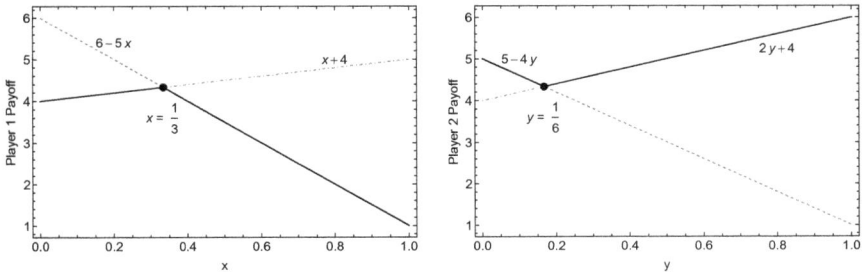

Fig. 5.6. Plotting the expected payoff to Bradley by playing the mixed strategy $\langle x, (1-x) \rangle$ when von Kluge plays pure strategies shows which strategy von Kluge should pick. When $x \leq 1/3$, von Kluge does better if he retreats because $x + 4$ is below $6 - 5x$. On the other hand, if $x \geq 1/3$, then von Kluge does better if he attacks because $6 - 5x$ is below $x + 4$. Remember, Von Kluge wants to *minimize* the payoff to Bradley. The point at which Bradley does best (i.e., maximizes his expected payoff) comes at $x = \frac{1}{3}$. By a similar argument, when $y \leq \frac{1}{6}$, Bradley does better if he chooses to move east (strategy one), while when $y \geq \frac{1}{6}$, Bradley does best when he waits (strategy two). Remember, Bradley is *minimizing* von Kluge's payoff since we are working with $-\mathbf{A}$.

strategy with $x \leq \frac{1}{3}$. On the other hand, if $x \geq 1/3$, then von Kluge does better if he attacks because $6 - 5x$ is below $x + 4$. That is, the best Bradley can hope to get is $6 - 5x$ if he announces to von Kluge that he is playing a mixed strategy with $x \geq \frac{1}{3}$. The break-even point occurs at $x = \frac{1}{3}$, at which it does not matter what strategy von Kluge plays.

By a similar argument, we can compute the expected payoff to von Kluge when he plays the mixed strategy $\langle y, (1-y) \rangle$ with $y \in [0, 1]$ and Bradley plays pure strategies. The expected payoff to von Kluge when Bradley plays strategy one is

$$\mathbf{e}_1^T(-\mathbf{A})\mathbf{y} = -y - 5(1 - y) = 4y - 5.$$

When Bradley plays strategy two, the expected payoff to von Kluge is

$$\mathbf{e}_2^T(-\mathbf{A})\mathbf{y} = -6y - 4(1 - y) = -2y - 4.$$

Reasoning that Bradley will attempt to minimize his payoff, von Kluge should consider the function

$$u_2(y) = \min\{4y - 5, -2y - 4\}$$

and find a value y that maximizes $u_2(y)$. Again, this maximizing point occurs at the intersection of the two lines or when $y = \frac{1}{6}$.

Note that we have done something subtle here. We are now treating both players as agents attempting to maximize their payoffs, rather than thinking of one agent as a minimizer and another agent as a maximizer. We take this approach when discussing computational methods for finding Nash equilibria in later chapters.

We now define

$$\mathbf{x}^* = \begin{bmatrix} \frac{1}{3} \\ \frac{2}{3} \end{bmatrix} \quad \text{and} \quad \mathbf{y}^* = \begin{bmatrix} \frac{1}{6} \\ \frac{5}{6} \end{bmatrix}.$$

The pair $(\mathbf{x}^*, \mathbf{y}^*)$ is the Nash equilibrium for this problem.

Example 5.66 (Saddle-Point Visualization). Continuing on from the previous example, we note that any Nash equilibrium for a zero-sum game is called a saddle point. To see why, consider the payoff function for Player 1 as a function of x and y (from the previous example). This function is

$$\begin{bmatrix} x & 1-x \end{bmatrix} \begin{bmatrix} 1 & 5 \\ 6 & 4 \end{bmatrix} \begin{bmatrix} y \\ 1-y \end{bmatrix} = 4 + x + 2y - 6xy. \tag{5.54}$$

We plot this in Fig. 5.7. The surface (with the corresponding contour plot) is a hyperbolic saddle. In three-dimensional space, it looks like a combination of an upside-down parabola going in one direction and a right-side-up parabola going in the other. The maximum of one parabola and the minimum of the other parabola occur precisely at the point $(x^*, y^*) = (\frac{1}{3}, \frac{1}{6})$. This is the point in the (x, y)-plane corresponding to the Nash equilibrium. We discuss this further in Section 6.7.

Remark 5.67. The techniques discussed in Examples 5.65 and 5.66 can be extended to cases when one player has two strategies and the other player has more than two strategies; however, these methods are not efficient for finding Nash equilibria in general. In the following chapters, we show how to find Nash equilibria for games by solving specific optimization problems corresponding to the games in question. These techniques will work for general two-player zero-sum games with an arbitrary number of strategies. We also discuss the problem of finding Nash equilibria in arbitrary bimatrix games.

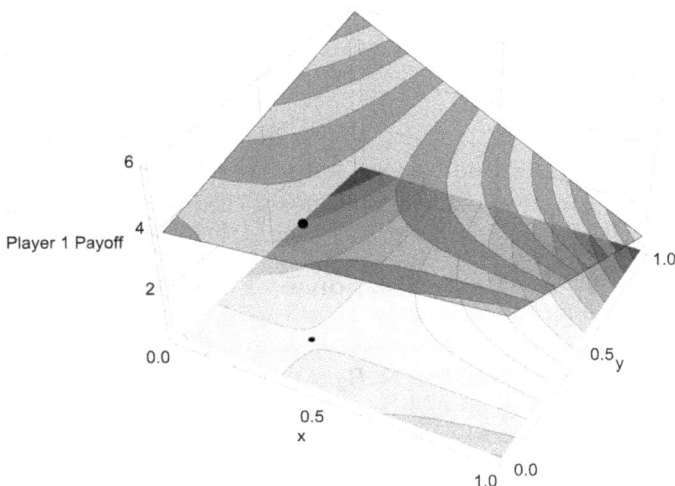

Fig. 5.7. The payoff function for Player 1 is shown as a function of x and y. *Note*: The Nash equilibrium occurs at a saddle point of the function.

5.9 Nash Equilibria in General-Sum Games

Remark 5.68. We now generalize our discussion on the existence of Nash equilibria to general-sum games with $N \geq 2$ players. The results in this section are (almost) completely analogous to those we have already seen, only that here they are in a more general setting. For the remainder of this section, all results will be about an N-player game, $\mathcal{G} = (\mathbf{P}, \Sigma, \pi)$, in normal form with $\Sigma_i = \{\sigma_1^i, \ldots, \sigma_{n_i}^i\}$, where Δ is the mixed-strategy space for this game.

Definition 5.69 (Player Best Response). If $\mathbf{y} = (\mathbf{y}^1, \ldots, \mathbf{y}^i, \ldots, \mathbf{y}^N) \in \Delta$ is a mixed strategy for all players, then the *best reply* for Player P_i is the set

$$B_i(\mathbf{y}) = \left\{ \mathbf{x}^i \in \Delta_{n_i} \colon u_i(\mathbf{x}^i, \mathbf{y}^{-i}) \geq u_i(\mathbf{z}^i, \mathbf{y}^{-i}) \quad \forall \mathbf{z}^i \in \Delta_{n_i} \right\}. \quad (5.55)$$

Recall that $\mathbf{y}^{-i} = (\mathbf{y}^1, \ldots, \mathbf{y}^{i-1}, \mathbf{y}^{i+1}, \ldots, \mathbf{y}^N)$ is the collection of mixed strategies not including Player i's.

Remark 5.70. There is a lot going on in Definition 5.69. What we are really saying is that, given some strategies, $\mathbf{y}^{-i} = (\mathbf{y}^1, \ldots, \mathbf{y}^{i-1}, \mathbf{y}^{i+1}, \ldots, \mathbf{y}^N)$, find all the strategies $\mathbf{x}^i \in \Delta_{n_i}$ for Player i that return the best possible payoffs, given all the other players are using $\mathbf{y}^{-i}$.

We can see that if a Player P_i is confronted by some collection of strategies $\mathbf{y}^{-i}$, then the best thing she can do is to choose some strategy $\mathbf{x}^i \in B_i(\mathbf{y})$.

Remark 5.71. Note that Eq. (5.55) defines the function $B_i: \Delta \to 2^{\Delta_{n_i}}$. That is, B_i is a point-to-set map. We can generalize this to the best response function, which combines these individual functions for each player.

Definition 5.72 (Best Response Function). The mapping $B: \Delta \to 2^{\Delta}$, given by

$$B(\mathbf{x}) = B_1(\mathbf{x}) \times B_2(\mathbf{x}) \times \cdots \times B_N(\mathbf{x}), \tag{5.56}$$

is called the *best response mapping*.

Theorem 5.73. *The strategy* $\mathbf{x}^* \in \Delta$ *is a Nash equilibrium for* $\mathcal{G}$ *if and only if* $\mathbf{x}^* \in B(\mathbf{x}^*)$.

Proof. Suppose that $\mathbf{x}$ is a Nash equilibrium. Then, for all $i \in \{1 \dots, N\}$,

$$u_i(\mathbf{x}^{i^*}, \mathbf{x}^{-i^*}) \geq u_i(\mathbf{z}^i, \mathbf{x}^{-i^*}),$$

for every $\mathbf{z}^i \in \Delta_{n_i}$. Thus,

$$\mathbf{x}^{i^*} \in \left\{ \mathbf{x}^i \in \Delta_{n_i} : u_i(\mathbf{x}^{i^*}, \mathbf{x}^{-i^*}) \geq u_i(\mathbf{z}^i, \mathbf{x}^{-i^*}) \quad \forall \mathbf{z}^i \in \Delta_{n_i} \right\}.$$

It follows that $\mathbf{x}^{i^*} \in B_i(\mathbf{x}^{i^*})$. Since this holds for each $i \in \{1, \dots, N\}$, it follows that $\mathbf{x}^* \in B(\mathbf{x}^*)$.

To prove the converse, suppose that $\mathbf{x}^* \in B(\mathbf{x}^*)$. Then, for all $i \in \{1, \dots, N\}$. Then, we have

$$\mathbf{x}^{i^*} \in \left\{ \mathbf{x}^i \in \Delta_{n_i} : u_i(\mathbf{x}^i, \mathbf{x}^{-i}) \geq u_i(\mathbf{z}^i, \mathbf{y}^{-i}) \quad \forall \mathbf{z}^i \in \Delta_{n_i} \right\}.$$

However, this implies that for all $i \in \{1 \dots, N\}$,

$$u_i(\mathbf{x}^{i^*}, \mathbf{x}^{-i^*}) \geq u_i(\mathbf{z}^i, \mathbf{x}^{-i^*}),$$

for every $\mathbf{z}^i \in \Delta_{n_i}$. This is the definition of a Nash equilibrium. $\square$

Remark 5.74. Theorem 5.73 shows that in the N-player, general-sum game setting, every Nash equilibrium is a kind of fixed point of the mapping $B: \Delta \to 2^\Delta$. This fact, along with a more general topological *fixed point theorem* called Kakutani's fixed point theorem, is sufficient to establish the following theorem.

Theorem 5.75 (Existence of Nash Equilibria). *Let $\mathcal{G} = (\mathbf{P}, \Sigma, \pi)$ be an N-player game in normal form. Then, $\mathcal{G}$ has at least one Nash equilibrium.*

Remark 5.76. The proof based on Kakutani's fixed point theorem is neither useful nor satisfying. Moreover, to apply Kakutani's theorem, we would have to prove that the mapping B has additional properties, which we have not discussed. Nash constructed an alternative proof using Brouwer's fixed point theorem, following the same steps we used to prove Theorem 5.63, which we now generalize.

Proof of the Existence of Nash Equilibria. Define the function

$$J_k^i(\mathbf{x}) = \begin{cases} u_i(\mathbf{e}_k, \mathbf{x}^{-i}) - u_i(\mathbf{x}^i, \mathbf{x}^{-i}) & \text{if this quantity is positive,} \\ 0 & \text{otherwise.} \end{cases}$$

(5.57)

This function measures the benefit of changing to the pure strategy $\mathbf{e}_k$ for Player P_i when all other players hold their strategy fixed at $\mathbf{x}^{-i}$. Now, define the transformation $T: \Delta \to \Delta$ so that

$$x_j^{i'} = \frac{x_j^i + J_j^i(\mathbf{x})}{1 + \sum_{k=1}^{n_i} J_k^i(\mathbf{x})}.$$

(5.58)

This is a generalization of Eqs. (5.50) and (5.51), and it follows from the same reasoning as in the proof of Theorem 5.63 that $x_1^{i'} + \cdots + x_{n_i}^{i'} = 1$ and $x_j^{i'} \geq 0$ for all i, thus establishing that T maps Δ to itself.

It now follows by the same reasoning as in Theorem 5.63 that $\mathbf{x}^*$ is a fixed point of T if and only if $\mathbf{x}^*$ is a Nash equilibrium. We assert that T is continuous by the continuity of the payoff function, and thus a fixed point for T exists by Brouwer's fixed point theorem. Consequently, every general-sum game has at least one Nash equilibrium. This completes the proof. $\square$

Remark 5.77. Unfortunately, this is still not a very useful way to construct a Nash equilibrium. In the following chapters, we explore this problem in depth for two-player zero-sum games and then proceed further to explore the problem for two-player general-sum games. The story of computing Nash equilibria takes on a life of its own. It is an important study within computational game theory and has had a substantial impact on the literature in mathematical programming (optimization), computer science, and economics.

5.10   Chapter Notes

John F. Nash studied at Princeton University, completing his thesis, *Non-Cooperative Games*, in 1950 and publishing it in 1951 [62]. During this time, Nash also worked on algebraic geometry, differential geometry, and partial differential equations. The influence these fields had on his work in game theory is clear from his use of the Kakutani and Brouwer fixed-point theorems. While he was awarded the Nobel Prize in Economics for his work on game theory, he is better remembered in some mathematical circles for his work in geometry and partial differential equations. Nash's embedding theorem [63, 64] shows that any Riemannian manifold (a fancy shape with special properties) can be isometrically embedded into some Euclidean space. Here, *isometric* means that lengths in the manifold are preserved during the process of embedding the shape in Euclidean space. By embedding, we mean a way of placing a shape inside a space. For example, imagine picking a circle up off a piece of paper and placing it in midair in front of you. This is a simple embedding of the circle into $\mathbb{R}^3$. Nash's approach to the proof worked by solving a system of partial differential equations and laid the foundations for what would become the Nash–Moser theorem [65, 66], which generalizes the inverse function theorem (see [67, Ch. 9]). His work, later recognized as the De Giorgi–Nash theorem, resolved Hilbert's 19th problem.

Thanks to the well-known biography *A Beautiful Mind* by Sylvia Nasar [68], it is widely known that Nash suffered from schizophrenia but slowly improved after 1970. The book was made into a movie of the same name. Interestingly, the depiction of Nash equilibrium in the movie is fantastically incorrect and is closer to describing Nash's bargaining theorem (which we discuss in Chapter 9).

$$- \spadesuit \clubsuit \heartsuit \diamondsuit -$$

5.11 Exercises

5.1 Show that the strategy $(\mathbf{e}_2, \mathbf{e}_1)$ is an equilibrium for the game in Example 5.3. That is, show that the strategy (drama, science fiction) is an equilibrium strategy for the networks.

5.2 Show that (Sail North, Search North) is an equilibrium solution for the Battle of the Bismark Sea using the approach from Example 5.3 and Theorem 5.2.

5.3 Prove Theorem 5.6.

5.4 Show that rock-paper-scissors does not have a saddle-point strategy.

5.5 Complete the proof of Proposition 5.29, and show explicitly that $u_2(\mathbf{x}, \mathbf{y}) = \mathbf{x}^T \mathbf{B} \mathbf{y}$.

5.6 Recall from Remark 5.33 that we wrote down what it meant for a Player 1 strategy to (weakly) dominate another in a two-player matrix game. Write these conditions for Player 2, assuming that we have the game $\mathcal{G} = (\mathbf{P}, \Sigma, \mathbf{A}, \mathbf{B})$. [Hint: Remember, Player 2 multiplies on the right-hand side of the payoff matrix. Also, you need to use $\mathbf{B}$.]

5.7 Show that confess strictly dominates don't confess for Clyde in Example 5.34 (prisoner's dilemma).

5.8 Using Theorem 5.37, state and prove an analogous theorem for Player 2.

5.9 Consider the game matrix (matrices) in Example 5.3. Show that this game is strictly dominance solvable. Recall that the game matrix is

$$\mathbf{A} = \begin{bmatrix} -15 & -35 & 10 \\ -5 & 8 & 0 \\ -12 & -36 & 20 \end{bmatrix}.$$

[Hint: Start with Player 2 (the column player) instead of Player 1. Note that column 3 is *strictly dominated* by column 1, so you can remove column 3. Proceed from there. You can eliminate two rows (or columns) at a time.]

5.10 Prove Lemma 5.55. [Hint: Argue that, for all $\mathbf{x} \in \Delta_m$ and for all $\mathbf{y} \in \Delta_n$, we know that $v_1(\mathbf{x}) \leq v_2(\mathbf{y})$ by showing that $v_2(\mathbf{y}) \geq \mathbf{x}^T \mathbf{A} \mathbf{y} \geq v_1(\mathbf{x})$. From this, conclude that $\min_{\mathbf{y}} v_2(\mathbf{y}) \geq \max_{\mathbf{x}} v_1(\mathbf{x})$.]

5.11 Suppose that $(\mathbf{x}^*, \mathbf{y}^*)$ is a Nash equilibrium for $\mathcal{G} = (\mathbf{P}, \Sigma, \mathbf{A}, \mathbf{B})$, with $\mathbf{A}, \mathbf{B} \in \mathbb{R}^{m \times n}$, and suppose $\mathcal{I}$ is a set of indices so that $i \in \mathcal{I}$ if and only if $x_i^* > 0$. Suppose that if $\mathbf{x} \in \Delta_m$ with the property that $x_i > 0$ if and only if $i \in \mathcal{I}$. Show that

$$\mathbf{x}^T \mathbf{A} \mathbf{y}^* = \mathbf{x}^{*T} \mathbf{A} \mathbf{y}^*,$$

thus generalizing the indifference theorem. [Hint: Use the indifference theorem, stating that $\mathbf{e}_i^T \mathbf{A} \mathbf{y}^* = \mathbf{x}^{*T} \mathbf{A} \mathbf{y}^*$ for all $i \in \mathcal{I}$ and the fact that $x_1 + \cdots x_m = 1$.]

5.12 Prove Lemma 5.62.

5.13 Consider the football game in Example 4.5. Compute the Nash equilibrium strategy. [Hint: Find the matrices $\mathbf{A}$ and $\mathbf{B} = -\mathbf{A}$ from the data. Use the method in Example 5.65 with Player 1, who has two strategies, and find $\mathbf{x}^* = \langle x^*, 1 - x^* \rangle$. Compute the row $\mathbf{x}^{*T} \mathbf{B}$. You will find that one strategy should be eliminated. This tells you the two remaining strategies to use. Now, apply the method in Example 5.65 to Player 2 and the two remaining strategies.]

5.14 Verify that the function T in Theorem 5.63 is continuous.

Part 2
Optimization and Game Theory

Chapter 6

An Introduction to Optimization and the Karush–Kuhn–Tucker Conditions

Chapter Goals: Many of the results we have encountered in game theory so far can be rephrased using the language of optimization. Our goal in this chapter is to introduce the basic elements of optimization theory necessary to derive optimization problems whose solutions will provide Nash equilibria for matrix games. Using a running example, we introduce the reader to the dual variables (called Lagrange multipliers in vector calculus), convex sets and functions, and the Karush–Kuhn–Tucker necessary conditions for optimality. Unlike in the previous chapters, we will not prove many of the theorems we state in this chapter, and instead take them as given. Detailed proofs for the results (or variations thereof) can be found in Ref. [69] or Ref. [70]. We will see in later chapters that these conditions can be used to derive the necessary (and sufficient) conditions for a pair of strategies to be a Nash equilibrium in a two-player matrix game.

6.1 Motivating Example

Example 6.1. We begin with a simple example that can be solved using techniques from an introductory calculus course. Suppose we wish to build a pen to keep some goats (or whatever farm animal you prefer). We are given 100 meters of fencing, and we wish to build the

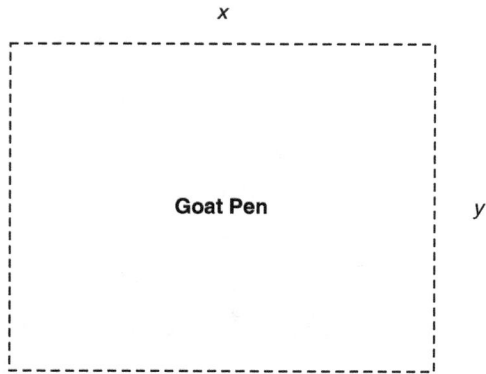

Fig. 6.1.   Goat pen with unknown side lengths: The objective is to identify the values of x and y that maximize the area of the pen (and thus the number of goats that can be kept).

pen in a rectangle with the largest possible area. How long should the sides of the rectangle be to maximize the resulting area?

The problem is illustrated in Fig. 6.1. Assume that we define the rectangle to have vertex points $(0,0)$, $(x,0)$, $(y,0)$, and (x,y), as illustrated in Fig. 6.1. We know that

$$2x + 2y = 100 \tag{6.1}$$

because $2x + 2y$ is the perimeter of the pen, and we are given 100 meters of fencing. The area of the pen is $A(x,y) = xy$. We can use Eq. (6.1) to solve for x in terms of y,

$$y = 50 - x; \tag{6.2}$$

therefore, $A(x) = x(50 - x)$. To maximize $A(x)$, recall that we take the first derivative of $A(x)$ with respect to x, set this derivative to zero, and solve for x:

$$\frac{dA}{dx} = 50 - 2x = 0. \tag{6.3}$$

Thus, $x = 25$ and $y = 50 - x = 25$. Recall from basic calculus that we can confirm that this is a maximum by evaluating the second derivative:

$$\left.\frac{d^2 A}{dx^2}\right|_{x=25} = -2 < 0. \tag{6.4}$$

The fact that this is negative implies that $x = 25$ is a *local maximum* for this function. Another way of seeing this is to note that $A(x) = 50x - x^2$ is an "upside-down" parabola. As we could have guessed, a square will maximize the area available for holding goats.

6.2 A General Maximization Formulation

Remark 6.2. We now generalize the goat pen example to study arbitrary optimization problems. The area function is a mapping from $\mathbb{R}^2$ to $\mathbb{R}$, written $A : \mathbb{R}^2 \to \mathbb{R}$. The domain of A is the two-dimensional space $\mathbb{R}^2$, and its range is $\mathbb{R}$. Our objective in Example 6.1 is to maximize the function A by choosing values for x and y. In optimization theory, the function we are trying to maximize (or minimize) is called the *objective function*. In general, an objective function is a mapping, $z : D \subseteq \mathbb{R}^n \to \mathbb{R}$. Here, D is the domain of the function z, usually taken to be all of $\mathbb{R}^n$.

Definition 6.3. Let $z : D \subseteq \mathbb{R}^n \to \mathbb{R}$. The point $\mathbf{x}^* \in D$ is a *global maximum* for z if for all $\mathbf{x} \in D$, $z(\mathbf{x}^*) \geq z(\mathbf{x})$. A point $\mathbf{x}^* \in D$ is a *local maximum* for z if there is a set $S \subseteq D$ with $\mathbf{x}^* \in S$ so that for all $\mathbf{x} \in S$, $z(\mathbf{x}^*) \geq z(\mathbf{x})$.

Remark 6.4. In Example 6.1, we are constrained in our choice of x and y by the fact that $2x + 2y = 100$. This is called a *constraint* in the optimization problem. More specifically, it's called an *equality constraint*. If we did not need to use all the fencing, then we could write the constraint as $2x + 2y \leq 100$, which is called an *inequality constraint*. In complex optimization problems, we can have many constraints. The set of all points in $\mathbb{R}^n$ for which the constraints are true is called the *feasible set* (or feasible region). In the goat pen problem, we wanted to find the best values of x and y to maximize the area $A(x, y)$. The variables x and y are called *decision variables*. We now generalize this.

Definition 6.5. Let $z : D \subseteq \mathbb{R}^n \to \mathbb{R}$ be a function. For $i \in \{1, \ldots, m\}$, let $g_i : D \subseteq \mathbb{R}^n \to \mathbb{R}$ be functions. Finally, for $j \in \{1, \ldots, l\}$, let $h_j : D \subseteq \mathbb{R}^n \to \mathbb{R}$ be functions. Then, the general maximization problem with objective function $z(x_1, \ldots, x_n)$, *inequality constraints* $g_i(x_1, \ldots, x_n) \leq b_i$ $(i \in \{1, \ldots, m\})$, and *equality constraints* $h_j(x_1, \ldots, x_n) = r_j$ $(j \in \{1, \ldots, l\})$ is written as

$$
\left\{
\begin{aligned}
&\max \ z(x_1, \ldots, x_n), \\
&\text{s.t. } g_1(x_1, \ldots, x_n) \leq b_1, \\
&\qquad\qquad \vdots \\
&\qquad g_m(x_1, \ldots, x_n) \leq b_m, \\
&\qquad h_1(x_1, \ldots, x_n) = r_1, \\
&\qquad\qquad \vdots \\
&\qquad h_l(x_1, \ldots, x_n) = r_l.
\end{aligned}
\right. \tag{6.5}
$$

Remark 6.6. Eq. (6.5) is also called a *mathematical programming problem*. Naturally, when constraints are involved, we define the global and local maxima for the objective function, $z(x_1, \ldots, x_n)$, in terms of the feasible region instead of the domain of D since we are only concerned with the values of $x_1, \ldots, x_n$ that satisfy the constraints.

Example 6.7 (Continuation of Example 6.1). We can rewrite the problem in Example 6.1 as

$$
\left\{
\begin{aligned}
&\max \ A(x, y) = xy, \\
&\text{s.t. } 2x + 2y \leq 100, \\
&\qquad x \geq 0, \\
&\qquad y \geq 0.
\end{aligned}
\right. \tag{6.6}
$$

Note that we have the option to change the constraint $2x + 2y = 100$ to the inequality constraint $2x + 2y \leq 100$. In reality, there is no requirement to use all the fencing, though that certainly seems prudent.

As a result, we've added two inequality constraints, $x \geq 0$ and $y \geq 0$, because it doesn't really make any sense to have negative lengths. We can rewrite these constraints as $-x \leq 0$ and $-y \leq 0$, where $g_1(x, y) = -x$ and $g_2(x, y) = -y$, to make Eq. (6.6) look like Eq. (6.5). We use this formulation of the problem in later examples.

Remark 6.8. We have formulated the general *maximization* problem in Eq. (6.5). Suppose that we are interested in finding a value that minimizes an objective function $z(x_1, \ldots, x_n)$, subject to certain constraints. Then, we can write Eq. (6.5), replacing max with min.

Remark 6.9. An alternative way of dealing with minimization is to transform a minimization problem into a maximization problem. If we want to minimize $z(x_1, \ldots, x_n)$, we can maximize $-z(x_1, \ldots, x_n)$. In maximizing the negation of the objective function, we are actually finding a value that minimizes $z(x_1, \ldots, x_n)$. This is particularly useful when dealing with code for optimization (e.g., SciPy's `linprog` at the time of this writing [71]).

6.3 Gradients, Constraints, and Optimization

Remark 6.10. We now make use of the relationship between the gradient of a function, its level sets, and optimization. The details needed for this are usually taught in a vector calculus class and can be found in Appendix B.

Definition 6.11 (Binding Constraint). Let $g(\mathbf{x}) \le b$ be a constraint in an optimization problem. If at point $\mathbf{x}_0 \in \mathbb{R}^n$ we have $g(\mathbf{x}_0) = b$, then the constraint is said to be *binding*. Equality constraints $h(\mathbf{x}) = r$ are always binding.

Example 6.12 (Continuation of Example 6.1). Recall Example 6.1. Consider the level curves of the objective function $z = xy$ and the constraint $2x + 2y \le 100$, which is binding at the optimal point $x = y = 25$. In Fig. 6.2, we see the level curves of the objective function (the hyperbolae) and the feasible region shown as a shaded triangle. The elements in the feasible regions are all values for x and y for which $2x + 2y \le 100$ and $x, y \ge 0$. Note that at the point of optimality, the level curve, defined by the equation

$$A(x, y) = xy = 625$$

is tangent to the line $2x + 2y = 100$. That is, the level curve of the objective function is tangent to the binding constraint.

The gradient of $A(x, y) = xy$ at this point is given by

$$\nabla A(25, 25) = \begin{bmatrix} y \\ x \end{bmatrix}\Bigg|_{x=25, y=25} = \begin{bmatrix} 25 \\ 25 \end{bmatrix}.$$

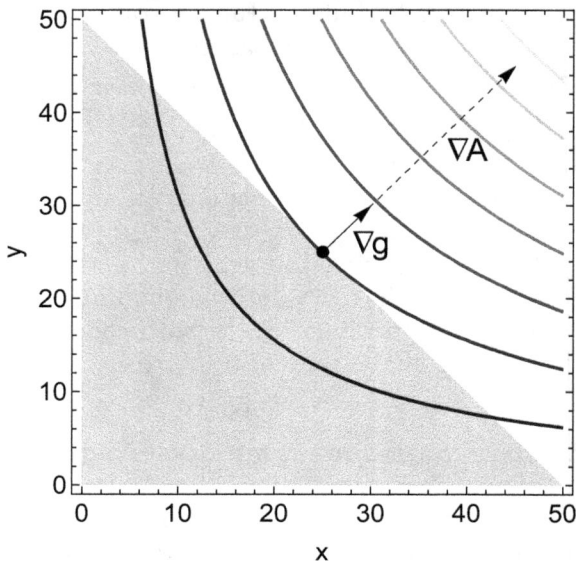

Fig. 6.2.   At optimality, the level curve of the objective function is tangent to the binding constraints.

We see that it is pointing in the direction of increase for the function $A(x, y)$, as expected. Now, let $g(x, y) = 2x + 2y$. Note that

$$\nabla g(25, 25) = \begin{bmatrix} 2 \\ 2 \end{bmatrix}.$$

This gradient is simply a scaled version of the gradient of the objective function. Let $(x^*, y^*) = (25, 25)$, and define $\lambda = \frac{25}{2}$ Then, we have

$$\nabla A(x^*, y^*) = \lambda \nabla g(x^*, y^*).$$

That is, the gradient of the objective function is simply a dilation of the gradient of the binding constraint. This is also illustrated in Fig. 6.2.

Remark 6.13. The elements illustrated in the previous example are true in general. You may recognize λ as a Lagrange multiplier from vector calculus [72]. We formalize this idea when we discuss Theorem 6.36, but first we need to divert and discuss a bit more geometry.

6.4   Convex Sets and Combinations

Remark 6.14. To understand optimization as it applies to the kinds of game theory we have been discussing, we need to introduce the concept of convexity.

Definition 6.15 (Convex Set). Let $X \subseteq \mathbb{R}^n$. The set X is convex if and only if, for all points $\mathbf{x}_1, \mathbf{x}_2 \in X$, we have $\lambda \mathbf{x}_1 + (1 - \lambda)\mathbf{x}_2 \in X$ for all $\lambda \in [0, 1]$.

Remark 6.16. This definition seems complex, but it is easy to understand. First, recall that if $\lambda \in [0, 1]$, then the point $\lambda \mathbf{x}_1 + (1 - \lambda)\mathbf{x}_2$ is on the line segment connecting $\mathbf{x}_1$ and $\mathbf{x}_2$ in $\mathbb{R}^n$. For example, when $\lambda = \frac{1}{2}$, then $\lambda \mathbf{x}_1 + (1 - \lambda)\mathbf{x}_2$ is the midpoint between $\mathbf{x}_1$ and $\mathbf{x}_2$. In fact, for every point $\mathbf{x}$ on the line connecting $\mathbf{x}_1$ and $\mathbf{x}_2$, we can find a value $\lambda \in [0, 1]$ so that $\mathbf{x} = \lambda \mathbf{x}_1 + (1 - \lambda)\mathbf{x}_2$. From this, we deduce that a set is convex if, given any pair of points $\mathbf{x}_1, \mathbf{x}_2 \in X$, the line segment connecting these points lies entirely inside X.

Example 6.17. Fig. 6.3 illustrates a convex and non-convex set. In two dimensions, non-convex sets have some pieces that resemble crescent shapes.

Theorem 6.18. *The intersection of a finite number of convex sets in $\mathbb{R}^n$ is convex.*

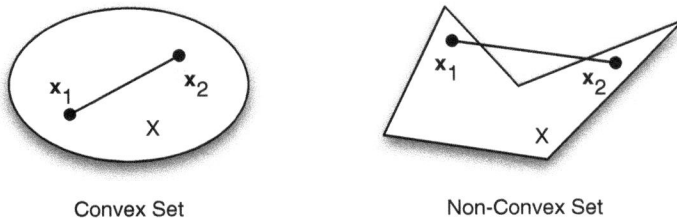

Convex Set   Non-Convex Set

Fig. 6.3.   The set on the left (an ellipse and its interior) is a convex set. Every pair of points inside the ellipse can be connected by a line contained entirely in the ellipse. The set on the right is clearly not convex. We have found two points whose connecting line is not contained inside the set.

Proof. Let $C_1, \ldots, C_n \subseteq \mathbb{R}^n$ be a finite collection of convex sets. Let

$$C = \bigcap_{i=1}^{n} C_i \tag{6.7}$$

be the set formed from the intersection of these sets. Choose $\mathbf{x}_1, \mathbf{x}_2 \in C$ and $\lambda \in [0, 1]$. Consider $\mathbf{x} = \lambda \mathbf{x}_1 + (1 - \lambda)\mathbf{x}_2$. We know that $\mathbf{x}_1, \mathbf{x}_2 \in C_1, \ldots, C_n$ by the definition of C. We know that $\mathbf{x} \in C_1, \ldots, C_n$ by the convexity of each set. Therefore, $\mathbf{x} \in C$. Thus, C is a convex set. $\qquad\square$

Definition 6.19 (Linear, Conical, and Convex Combinations). Let $\mathbf{x}_1, \ldots, \mathbf{x}_m$ be vectors in $\in \mathbb{R}^n$, and let $\alpha_1, \ldots, \alpha_m \in \mathbb{R}$ be scalars. Then,

$$\mathbf{y} = \alpha_1 \mathbf{x}_1 + \cdots + \alpha_m \mathbf{x}_m \tag{6.8}$$

is a *linear combination* of the vectors $\mathbf{x}_1, \ldots, \mathbf{x}_m$.

If $\alpha_1, \ldots, \alpha_m \geq 0$, then Eq. (6.8) is called a *conical combination* (or sometimes a non-negative combination) of the vectors. Moreover, if $\alpha_1, \ldots, \alpha_m \in [0, 1]$ and

$$\alpha_1 + \cdots + \alpha_m = 1,$$

then Eq. (6.8) is called a *convex combination* of the vectors. If $\alpha_i < 1$ for all $i \in \{1, \ldots, m\}$, then Eq. (6.8) is called a *strict convex combination*.

Remark 6.20. We can see that we move from the very general to the very specific as we go from linear combinations to conical combinations to convex combinations. A linear combination of points or vectors allows us to choose any real value for the coefficients. A conical combination restricts us to positive values, while a convex combination asserts that those values must be non-negative and sum to 1.

Theorem 6.21. *Let $\mathbf{x}_1, \ldots, \mathbf{x}_m$ be vectors in $\mathbb{R}^n$. Then, the set*

$$\mathrm{Cone}(\mathbf{x}_1, \ldots, \mathbf{x}_n) = \left\{ \mathbf{y} \in \mathbb{R}^n : \mathbf{y} = \sum_i \alpha_i \mathbf{x}_i, \ \alpha_i \geq 0 \right\}$$

is convex.

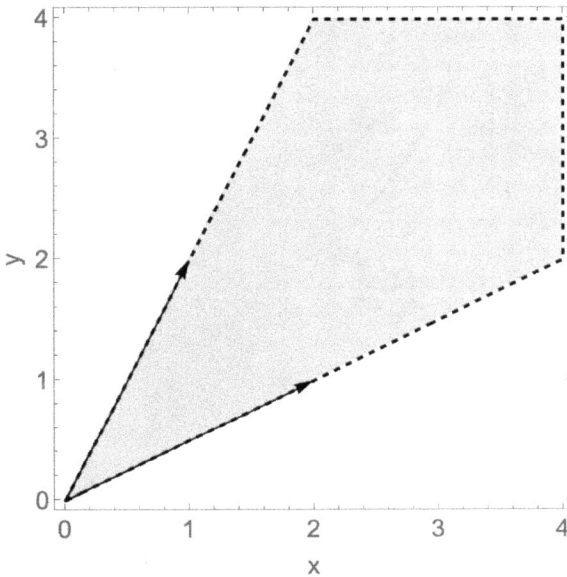

Fig. 6.4. The cone generated from the vectors $\mathbf{x}_1 = \langle 1, 2 \rangle$ and $\mathbf{x}_2 = \langle 2, 1 \rangle$. This looks like a squashed version of the cone discussed in elementary school.

Remark 6.22. The set $\text{Cone}(\mathbf{x}_1, \ldots, \mathbf{x}_n)$ is called the *cone* or *conical hull*, generated by the vectors $\mathbf{x}_1, \ldots, \mathbf{x}_n$. This geometric structure appears regularly in the study of optimization.

Example 6.23. The cone generated by the vectors $\mathbf{x}_1 = \langle 1, 2 \rangle$ and $\mathbf{x}_2 = \langle 2, 1 \rangle$ is shown in Fig. 6.4. It is easy to see through visual inspection that this set is convex. In three dimensions, a cone looks more like the "cones" we are familiar with from elementary school or vector calculus.

6.5   Convex and Concave Functions

Definition 6.24 (Convex Function). A function $f : \mathbb{R}^n \to \mathbb{R}$ is convex if it satisfies

$$f(\lambda \mathbf{x}_1 + (1 - \lambda)\mathbf{x}_2) \leq \lambda f(\mathbf{x}_1) + (1 - \lambda)f(\mathbf{x}_2) \tag{6.9}$$

for all $\mathbf{x}_1, \mathbf{x}_2 \in \mathbb{R}^n$ and for all $\lambda \in [0, 1]$. The function is strictly convex if the inequality is replaced with a strict inequality.

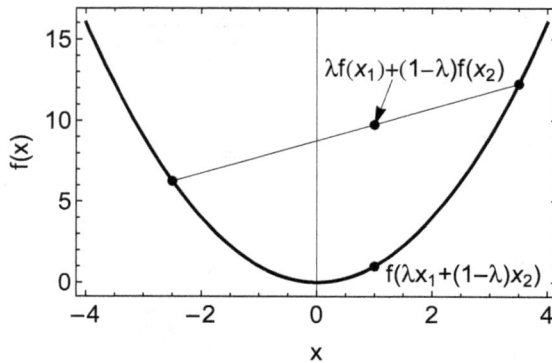

Fig. 6.5. A convex function satisfies the expression $f(\lambda x_1 + (1 - \lambda)x_2) \leq \lambda f(x_1) + (1 - \lambda)f(x_2)$ for all x_1 and x_2 and $\lambda \in [0, 1]$.

Example 6.25. A convex function is illustrated in Fig. 6.5. Here, we see that the graph of the function lies below all of its secant lines. Recall that a secant line connects two points on the graph of the function.

Definition 6.26 (Concave Function). A function $f : \mathbb{R}^n \to \mathbb{R}$ is convex if it satisfies

$$f(\lambda x_1 + (1 - \lambda)x_2) \geq \lambda f(x_1) + (1 - \lambda)f(x_2) \qquad (6.10)$$

for all $x_1, x_2 \in \mathbb{R}^n$ and for all $\lambda \in [0, 1]$. The function is strictly concave if the inequality is replaced with a strict inequality.

Remark 6.27. To visualize this definition, simply flip Fig. 6.5 upside down. The graphs of concave functions lie above their secant lines. In calculus, these functions are called "concave down," while convex functions are called "concave up." Those terms are not used outside of calculus, and it is best to ignore (or forget) them in favor of the terms convex and concave.

Remark 6.28. The following theorem relates convex functions and convex sets. Its proof is outside the scope of this book but can be found in Ref. [69]. Note that there is no such thing as a concave set.

Theorem 6.29. *Let $f : \mathbb{R}^n \to \mathbb{R}$ be a convex function. Then, the set $C = \{x \in \mathbb{R}^n : f(x) \leq c\}$, where $c \in \mathbb{R}$, is a convex set.*

Definition 6.30 (Linear Function). A function $g : \mathbb{R}^n \to \mathbb{R}$ is *linear* if there are constants $c_1, \dots, c_n \in \mathbb{R}$ so that

$$g(x_1, \dots, x_n) = c_1 x_1 + \cdots + c_n x_n. \tag{6.11}$$

Example 6.31. We have had experience with many linear functions already. The left-hand side of the constraint $2x + 2y \leq 100$ is a linear function. That is, the function $g(x, y) = 2x + 2y$ is a linear function of x and y.

Remark 6.32. It is worth noting that linear functions can be defined in a much more general context. This is usually handled in a linear algebra course. See Ref. [73] for details. For us, Definition 6.30 is more than sufficient.

Definition 6.33 (Affine Function). A function $g : \mathbb{R}^n \to \mathbb{R}$ is *affine* if $g(\mathbf{x}) = l(\mathbf{x}) + b$, where $l : \mathbb{R}^n \to \mathbb{R}$ is a linear function and $b \in \mathbb{R}$.

Remark 6.34. It is easy to see that every linear function is affine by setting $b = 0$. While linear functions are easier to deal with, it is often easier to state theorems in terms of affine functions, especially when these affine functions appear as constraints.

6.6 Karush–Kuhn–Tucker Conditions

Remark 6.35. We now make use of all the machinery we have constructed to state the Karush–Kuhn–Tucker (KKT) theorem. This powerful theorem provides the necessary conditions for when a point $\mathbf{x}^* \in \mathbb{R}^n$ will maximize an objective function $z(\mathbf{x})$, subject to some constraints given by additional functions. We state this theorem (and its corollaries) but do not prove them. Proofs can be found in Ref. [69, 70].

Theorem 6.36. *Let $z : \mathbb{R}^n \to \mathbb{R}$ be a differentiable objective function, $g_i : \mathbb{R}^n \to \mathbb{R}$ be differentiable constraint functions for $i \in \{1, \dots, m\}$, and $h_j : \mathbb{R}^n \to \mathbb{R}$ be differentiable constraint functions* *for $j \in \{1, \ldots, l\}$. If $\mathbf{x}^* \in \mathbb{R}^n$ is an optimal point satisfying an appropriate regularity condition for the following optimization problem,*

$$
P \begin{cases}
\text{max } z(x_1, \ldots, x_n), \\
\quad s.t. \ g_1(x_1, \ldots, x_n) \leq 0, \\
\qquad \vdots \\
\qquad g_m(x_1, \ldots, x_n) \leq 0, \\
\qquad h_1(x_1, \ldots, x_n) = 0, \\
\qquad \vdots \\
\qquad h_l(x_1, \ldots, x_n) = 0,
\end{cases}
$$

then there are constants, $\lambda_1, \ldots, \lambda_m \in \mathbb{R}$ and $\mu_1, \ldots \mu_l \in \mathbb{R}$, so that we have the following:

$$
\textit{Primal feasibility}: \begin{cases}
g_i(\mathbf{x}^*) \leq 0 & \text{for } i \in \{1, \ldots, m\}, \\
h_j(\mathbf{x}^*) = 0 & \text{for } j \in \{1, \ldots, l\}.
\end{cases}
$$

$$
\textit{Dual feasibility}: \begin{cases}
\nabla z(\mathbf{x}^*) - \displaystyle\sum_{i=1}^{m} \lambda_i \nabla g_i(\mathbf{x}^*) - \sum_{j=1}^{l} \mu_j \nabla h_j(\mathbf{x}^*) = \mathbf{0}, \\
\qquad\qquad\quad \lambda_i \geq 0 \quad \text{for } i \in \{1, \ldots, m\}, \\
\qquad\qquad\quad \mu_j \in \mathbb{R} \quad \text{for } j \in \{1, \ldots, l\}.
\end{cases}
$$

$$
\textit{Complementary slackness}: \{\ \lambda_i g_i(\mathbf{x}^*) = 0 \quad \text{for } i \in \{1, \ldots, m\}.
$$

Remark 6.37. The regularity condition mentioned in Theorem 6.36 is sometimes called a constraint qualification. A common one is that the gradients of the binding constraints are all linearly independent at $\mathbf{x}^*$. There are many variations of constraint qualifications. These are detailed in texts on optimization [69, 70]. Suffice it to say that all the problems we consider will automatically satisfy a constraint qualification, meaning that the KKT theorem holds.

Remark 6.38. The KKT theorem provides the *necessary* conditions for a point to be an optimal solution to a constrained maximization problem. The following theorem provides a sufficient condition for a point to be a global optimal solution to a constrained maximization problem. You will note that the conditions are nearly identical – it is the order of implication that has changed, as has the imposition of concavity and convexity requirements on the functions involved in the problem.

Theorem 6.39. *Let $z : \mathbb{R}^n \to \mathbb{R}$ be a differentiable concave function, $g_i : \mathbb{R}^n \to \mathbb{R}$ be differentiable convex functions for $i \in \{1, \ldots, m\}$, and $h_j : \mathbb{R}^n \to \mathbb{R}$ be affine functions for $j \in \{1, \ldots, l\}$. Suppose there are constants $\lambda_1, \ldots, \lambda_m \in \mathbb{R}$ and $\mu_1, \ldots \mu_l \in \mathbb{R}$ so that we have the following:*

$$\textit{Primal feasibility}: \begin{cases} g_i(\mathbf{x}^*) \leq 0 & \textit{for } i \in \{1, \ldots, m\}, \\ h_j(\mathbf{x}^*) = 0 & \textit{for } j \in \{1, \ldots, l\}. \end{cases}$$

$$\textit{Dual feasibility}: \begin{cases} \nabla z(\mathbf{x}^*) - \displaystyle\sum_{i=1}^{m} \lambda_i \nabla g_i(\mathbf{x}^*) - \sum_{j=1}^{l} \mu_j \nabla h_j(\mathbf{x}^*) = \mathbf{0}, \\ \\ \lambda_i \geq 0 \quad \textit{for } i \in \{1, \ldots, m\}, \\ \mu_j \in \mathbb{R} \quad \textit{for } j \in \{1, \ldots, l\}. \end{cases}$$

$$\textit{Complementary slackness}: \{\, \lambda_i g_i(\mathbf{x}^*) = 0 \quad \textit{for } i \in \{1, \ldots, m\}.$$

Then, $\mathbf{x}^*$ is a global maximum for the problem

$$P \begin{cases} \max \ z(x_1, \ldots, x_n), \\ \quad s.t. \ g_1(x_1, \ldots, x_n) \leq 0, \\ \qquad\qquad \vdots \\ \qquad g_m(x_1, \ldots, x_n) \leq 0, \\ \qquad h_1(x_1, \ldots, x_n) = 0, \\ \qquad\qquad \vdots \\ \qquad h_l(x_1, \ldots, x_n) = 0. \end{cases}$$

Remark 6.40. The values $\lambda_1, \ldots, \lambda_m$ and $\mu_1, \ldots, \mu_l$ are sometimes called *Lagrange multipliers* or *dual variables*. Primal feasibility, dual feasibility, and complementary slackness are called the KKT conditions.

Remark 6.41. Note that Theorem 6.36 is true even if $z(\mathbf{x})$ is not concave, the functions $g_i(\mathbf{x})$ ($i \in \{1, \ldots, m\}$) are not convex, or the functions $h_j(\mathbf{x})$ ($j \in \{1, \ldots, l\}$) are not linear. This is because it is the direction of the implication in Theorem 6.36 that matters. In particular, the fact that a triple, $(\mathbf{x}, \boldsymbol{\lambda}, \boldsymbol{\mu}) \in \mathbb{R}^n \times \mathbb{R}^m \times \mathbb{R}^l$, satisfies the KKT conditions does not ensure that this is an optimal solution for problem P. However, finding such triples and then evaluating the objective function at $\mathbf{x}^*$ is one way to solve optimization problems.

Remark 6.42. Looking more closely at the dual feasibility conditions, we see something interesting. Suppose that there are *no* equality constraints, i.e., no constraints of the form $h_j(\mathbf{x}) = 0$. Then, the statements

$$\nabla z(\mathbf{x}^*) - \sum_{i=1}^{m} \lambda_i \nabla g_i(\mathbf{x}^*) - \sum_{j=1}^{l} \mu_j \nabla h_j(\mathbf{x}^*) = \mathbf{0},$$

$$\lambda_i \geq 0 \quad \text{for } i \in \{1, \ldots, m\}$$

imply that

$$\nabla z(\mathbf{x}^*) = \sum_{i=1}^{m} \lambda_i \nabla g_i(\mathbf{x}^*),$$

$$\lambda_i \geq 0 \quad \text{for } i \in \{1, \ldots, m\}.$$

Specifically, this says that the *gradient of z at* $\mathbf{x}^*$ *is a conical combination of the gradients of the constraints at* $\mathbf{x}^*$. But more importantly, since we also have *complementary slackness*, we know that if $\mathbf{g}_i(\mathbf{x}^*) \neq 0$, then $\lambda_i = 0$ because $\lambda_i g_i(\mathbf{x}^*) = 0$ for $i = 1, \ldots, m$. Thus, what dual feasibility is really saying is that the *gradient of z at* $\mathbf{x}^*$ *is a conical combination of the gradients of the* **binding** *constraints at* $\mathbf{x}^*$. Remember that a constraint is binding if $g_i(\mathbf{x}^*) = 0$, in which case $\lambda_i \geq 0$.

Remark 6.43. Continuing from the previous remark, in the general case when we have some equality constraints, then dual feasibility states that

$$\nabla z(\mathbf{x}^*) = \sum_{i=1}^{m} \lambda_i \nabla g_i(\mathbf{x}^*) + \sum_{j=1}^{l} \mu_j \nabla h_j(\mathbf{x}^*),$$

$$\lambda_i \geq 0 \quad \text{for } i \in \{1, \ldots, m\},$$

$$\mu_j \in \mathbb{R} \quad \text{for } j \in \{1, \ldots, l\}.$$

Since equality constraints are *always binding*, this states that the *gradient of z at* $\mathbf{x}^*$ *is a linear combination of the gradients of the* **binding** *constraints at* $\mathbf{x}^*$.

Example 6.44. We now come full circle and revisit Example 6.1. First, we reformulate the problem to match the style of Theorem 6.36:

$$\left\{\begin{array}{l} \max \ A(x,y) = xy, \\ \text{s.t.} \ 2x + 2y - 100 = 0, \\ \qquad -x \leq 0, \\ \qquad -y \leq 0. \end{array}\right. \tag{6.12}$$

Note that the greater-than inequalities $x \geq 0$ and $y \geq 0$ in Eq. (6.6) have been changed to less-than inequalities by multiplying them by -1. The constraint $2x + 2y = 100$ has simply been transformed to $2x + 2y - 100 = 0$. Thus, if $h(x,y) = 2x + 2y - 100$, we can see that $h(x,y) = 0$ is our constraint. We can let $g_1(x,y) = -x$ and $g_2(x,y) = -y$. Then, we have $g_1(x,y) \leq 0$ and $g_2(x,y) \leq 0$ as our inequality constraints. We already know that $x = y = 25$ is our optimal solution. Thus, we know that there must be Lagrange multipliers μ, λ_1, and λ_2 corresponding to the constraints $h(x,y) = 0$, $g_1(x,y) \leq 0$, and $g_2(x,y) \leq 0$ that satisfy the KKT conditions.

Let's investigate the three components of the KKT conditions:

Primal feasibility. If $x = y = 25$, then $h(25, 25) = 0$, $g_1(25, 25) = -25 \leq 0$, and $g_2(25, 25) = -25 \leq 0$. So, primal feasibility is satisfied.

Complementary slackness. We know that $g_1(x,y) = g_2(x,y) = -25$. Since neither of these functions is 0, we know that $\lambda_1 = \lambda_2 = 0$. This will force the conditions of complementary slackness to hold, namely,

$$\lambda_1 g_1(25, 25) = 0,$$
$$\lambda_2 g_2(25, 25) = 0.$$

Dual feasibility. We now know that $\lambda_1 = \lambda_2 = 0$. That means we need to find $\mu \in \mathbb{R}$ so that

$$\nabla A(25, 25) - \mu \nabla h(25, 25) = \mathbf{0}.$$

We know that

$$\nabla A(x, y) = \nabla(xy) = \begin{bmatrix} y \\ x \end{bmatrix},$$

$$\nabla h(x, y) = \nabla(2x + 2y - 100) = \begin{bmatrix} 2 \\ 2 \end{bmatrix}.$$

Evaluating $\nabla A(25, 25)$ yields

$$\begin{bmatrix} 25 \\ 25 \end{bmatrix} - \mu \begin{bmatrix} 2 \\ 2 \end{bmatrix} = \begin{bmatrix} 0 \\ 0 \end{bmatrix}.$$

Thus, setting $\mu = \frac{25}{2}$ will accomplish our goal.

We have identified the Lagrange multipliers $\lambda_1 = 0$, $\lambda_2 = 0$ and $\mu = \frac{25}{2}$ corresponding to the optimal point $x = y = 25$.

6.7 Relating Back to Game Theory

Remark 6.45. Game theory and optimization theory are intimately related. When you play a game, you're trying to maximize your payoff, subject to constraints on your moves *and* to the actions of the other players. We can formalize this idea.

Remark 6.46. Consider a game $\mathcal{G} = (\mathbf{P}, \Sigma, \pi)$ in normal form. Assume that $\mathbf{P} = \{P_1, \ldots, P_N$ and $\Sigma_i = \{\sigma_1^i, \ldots, \sigma_{n_i}^i\}$. If we assume a fixed mixed strategy $\mathbf{x} \in \Delta$, Player P_i's objective when choosing a response $\mathbf{x}^i \in \Delta_{n_i}$ is to solve the following problem:

$$\text{Player } P_i : \begin{cases} \max\ u_i(\mathbf{x}^i, \mathbf{x}^{-i}), \\ s.t.\ \mathbf{x}_1^i + \cdots + \mathbf{x}_{n_i}^i = 1, \\ \quad \mathbf{x}_j^i \geq 0 \quad j \in \{1, \ldots, n_i\}. \end{cases} \tag{6.13}$$

Recall that $\mathbf{x}^{-i}$ are the strategies for all the players except for Player i. Eq. (6.13) is a mathematical programming problem, provided that $u_i(\mathbf{x}^i, \mathbf{x}^{-i})$ is known. This specific optimization problem assumes that all players, except for Player i, are holding their strategy constant, e.g., playing $\mathbf{x}^{-i}$. In reality, each player is solving a version of this problem *simultaneously*. Thus, an equilibrium solution solves the following simultaneous system of optimization problems:

$$\forall i : \begin{cases} \max \ u_i(\mathbf{x}^i, \mathbf{x}^{-i}), \\ s.t. \ \mathbf{x}_1^i + \cdots + \mathbf{x}_{n_i}^i = 1, \\ \mathbf{x}_j^i \geq 0 \quad j \in \{1, \ldots, n_i\}. \end{cases} \tag{6.14}$$

This formulation leads to a rich class of problems in mathematical programming, which we discuss in the following chapters. However, we can easily illustrate this for a two-strategy game.

Example 6.47 (Finding the Nash equilibrium of the Battle of Avranches). Consider the Player 1 game matrix from the Battle of Avranches:

$$\mathbf{A} = \begin{bmatrix} 1 & 5 \\ 6 & 4 \end{bmatrix}.$$

Let $\mathbf{x} = \langle x, 1 - x \rangle$ and $\mathbf{y} = \langle y, 1 - y \rangle$, where $x, y \in [0, 1]$. We can construct the payoff function for Player 1 as

$$u_1(x, y) = \begin{bmatrix} x & 1 - x \end{bmatrix} \begin{bmatrix} 1 & 5 \\ 6 & 4 \end{bmatrix} \begin{bmatrix} y \\ 1 - y \end{bmatrix} = 4 + x + 2y - 6xy. \tag{6.15}$$

Recall that Player 1 wishes to maximize this function, while Player 2 wishes to minimize this function because this is a zero-sum game, where the gains of Player 1 are the losses of Player 2, and vice versa. For a moment, we ignore the constraints that $0 \leq x \leq 1$ and $0 \leq y \leq 1$ and simply compute the necessary conditions for optimality. In this case, we have

$$\text{Player 1} \begin{cases} \dfrac{\partial u_1}{\partial x} = 1 - 6y = 0, \end{cases}$$

$$\text{Player 2} \begin{cases} \dfrac{\partial u_1}{\partial y} = 2 - 6x = 0. \end{cases}$$

Note that this could simply be written as

$$\nabla u_1 = \mathbf{0},$$

where ∇ is computed in terms of x and y.

We can solve the equations for Player 1 and Player 2 to obtain $x^* = \frac{1}{3}$ and $y^* = \frac{1}{6}$. This yields a candidate Nash equilibrium solution:

$$\mathbf{x}^* = \begin{bmatrix} \frac{1}{3} \\ \frac{2}{3} \end{bmatrix}, \quad \mathbf{y}^* = \begin{bmatrix} \frac{1}{6} \\ \frac{5}{6} \end{bmatrix}.$$

One can easily verify whether this is a Nash equilibrium by letting $\langle x, 1 - x \rangle$ and $\langle y, 1 - y \rangle$ be arbitrary strategies for Player 1 and Player 2, respectively, and computing that

$$\frac{13}{3} = \begin{bmatrix} \frac{1}{3} & \frac{2}{3} \end{bmatrix} \begin{bmatrix} 1 & 5 \\ 6 & 4 \end{bmatrix} \begin{bmatrix} \frac{1}{6} \\ \frac{5}{6} \end{bmatrix} \geq \begin{bmatrix} x & 1-x \end{bmatrix} \begin{bmatrix} 1 & 5 \\ 6 & 4 \end{bmatrix} \begin{bmatrix} \frac{1}{6} \\ \frac{5}{6} \end{bmatrix} = \frac{13}{3},$$

$$\frac{13}{3} = \begin{bmatrix} \frac{1}{3} & \frac{2}{3} \end{bmatrix} \begin{bmatrix} 1 & 5 \\ 6 & 4 \end{bmatrix} \begin{bmatrix} \frac{1}{6} \\ \frac{5}{6} \end{bmatrix} \leq \begin{bmatrix} \frac{1}{3} & \frac{2}{3} \end{bmatrix} \begin{bmatrix} 1 & 5 \\ 6 & 4 \end{bmatrix} \begin{bmatrix} y \\ 1-y \end{bmatrix} = \frac{13}{3}.$$

Thus, neither player has an incentive to change strategies,[1] and so the strategy we identified must be a Nash equilibrium.

We saw this solution visualized in Fig. 5.7. The equilibrium occurs at a saddle point of the function. This is a point that is simultaneously a minimum in one direction and a maximum in another direction, justifying the notion that one player is pushing the value of the payoff down and the other player is pushing its value up.

6.8 Chapter Notes

It has only been in the past few decades that the KKT theorem has been named after Karush, Kuhn, and Tucker. Before this, the theorem was named after only Harold Kuhn and Albert Tucker [74]. It was rediscovered [75] that William Karush, in fact, published a version of the theorem in 1939 as part of his master's thesis [76], though Kuhn and Tucker did not try to hide this and were aware of the prior work in their 1951 publication. The theorem has subsequently been renamed the Karush–Kuhn–Tucker theorem in popular use, though

---

[1]You can also use Nash's map, $T : \Delta \to \Delta$, to show that this point is a Nash equilibrium, if you like.

older texts will still refer to it as the Kuhn–Tucker theorem. Interestingly, Karush worked for the Manhattan Project during World War II, ultimately becoming a devout advocate for peace [75].

Harold W. Kuhn was a mathematician and colleague of John Nash (after whom the Nash equilibrium is named) and was instrumental in bringing Nash's work to the attention of the Nobel Prize committee [77]. In addition to working with Albert W. Tucker (Nash's advisor) [77], he is also known for his general work on optimization, including the "Hungarian Algorithm" [78], which provides a solution to the assignment problem and can be used for various practical purposes.

There are ways to generalize the various theorems presented in this chapter. The most common way is to consider generalizations of convex functions to "quasi-convex" or "pseudo-convex" functions [70]. These generalizations still allow the KKT conditions to be sufficient for an optimal solution. Convex optimization is a considerable field of study in itself. Boyd and Vandenberghe's text on the subject provides a thorough starting point [79].

From a practical perspective, optimization is one of the three primary pillars of machine learning [80] (the other two being probability theory and linear algebra, for use in optimization). Prior to this, substantial work in numerical optimization (i.e., finding optimal solutions to real-world problems) has resulted in several algorithms for identifying solutions to optimization problems. Nocedal and Wright provide a thorough introduction to this topic [81]. In the coming chapters, we exploit the theory of optimization (in the form of the KKT conditions), along with existing numerical algorithms, to find methods for solving matrix games.

$$- \spadesuit \clubsuit \heartsuit \diamondsuit -$$

6.9 Exercises

6.1 Find the radius and height of a cylinder that has minimal surface area, assuming that its volume must be 27 cubic units. [Hint: Suppose that the can has radius r and height h. The formula for the surface area of a cylinder is $2\pi r h + 2\pi r^2$. The volume of a cylinder is $\pi r^2 h$ and is constrained to be equal to 27.]

6.2   Write the problem from Exercise 6.1 as a general minimization problem. Add any appropriate non-negativity constraints. [Hint: You must change max to min.]

6.3   Plot the level sets of the objective function and the feasible region in Exercise 6.1. At the point of optimality you identified, show that the gradient of the objective function is a scaled version of the gradient (linear combination) of the binding constraints.

6.4   Find the values of the dual variables for the optimal point in Exercise 6.1. Show that the KKT conditions hold for the values you found.

6.5   Using analogous reasoning, write the definitions for global and local minima. [Hint: Think about what a minimum means and find the correct direction for the $\geq$ sign in the definition above.]

6.6   Prove Theorem 6.29.

6.7   Prove that every affine function is both convex and concave.

Chapter 7

Linear Programming and Zero-Sum Games

Chapter Goals: In this chapter, we show how to convert any two-player zero-sum matrix game into an optimization problem that has a linear objective function and linear constraints. Such a problem is called a *linear programming problem* (for historical reasons). Readily available computer software can then be used to find the Nash equilibria as optimal solutions to these optimization problems. Thus, we provide an algorithmic way to solve a two-player zero-sum game with any number of strategies. To accomplish this, we explore a special property of linear programming problems, namely the fact that each linear programming problem has a partner linear programming problem called its dual. These two problems will provide the Nash equilibria for the two players in a zero-sum matrix game.

7.1   Linear Programs

Definition 7.1. A linear programming problem is an optimization problem with the following form:

$$
\begin{cases}
\max \ z(x_1, \ldots, x_n) = c_1 x_1 + \cdots + c_n x_n, \\
s.t. \ a_{11} x_1 + \cdots + a_{1n} x_n \leq b_1, \\
\qquad \vdots \\
\quad a_{m1} x_1 + \cdots + a_{mn} x_n \leq b_m, \\
\quad h_{11} x_1 + \cdots + h_{n1} x_n = r_1, \\
\qquad \vdots \\
\quad h_{l1} x_1 + \cdots + h_{ln} x_n = r_l.
\end{cases}
\tag{7.1}
$$

Here, a_{ij}, h_{kj}, b_i, and r_k are all real numbers for $i \in \{1, \ldots, m\}$, $k \in \{1, \ldots, l\}$, and $j \in \{1, \ldots, n\}$.

Remark 7.2. We can use matrices to write these problems more compactly. Consider the following system of equations:

$$
\begin{cases}
a_{11} x_1 + a_{12} x_2 + \cdots + a_{1n} x_n = b_1, \\
a_{21} x_1 + a_{22} x_2 + \cdots + a_{2n} x_n = b_2, \\
\qquad \vdots \\
a_{m1} x_1 + a_{m2} x_2 + \cdots + a_{mn} x_n = b_m.
\end{cases}
\tag{7.2}
$$

Then, we can write this in matrix notation as

$$
\mathbf{A x} = \mathbf{b},
\tag{7.3}
$$

where $\mathbf{A}_{ij} = a_{ij}$ for $i \in \{1, \ldots, m\}$, $j \in \{1, \ldots, n\}$, $\mathbf{x}$ is a column vector in $\mathbb{R}^n$ with entries x_j for $j \in \{1, \ldots, n\}$, and $\mathbf{b}$ is a column vector in $\mathbb{R}^m$ with entries b_i for $i \in \{1 \ldots, m\}$. If we replace the equalities in Eq. (7.3) with inequalities, we can also express systems of inequalities in the form

$$
\mathbf{A x} \leq \mathbf{b}.
\tag{7.4}
$$

Using this representation, we can write our general linear programming problem using matrix and vector notation. Equation (7.1) becomes

$$\begin{cases} \max & z(\mathbf{x}) = \mathbf{c}^T\mathbf{x}, \\ \text{s.t.} & \mathbf{A}\mathbf{x} \le \mathbf{b}, \\ & \mathbf{H}\mathbf{x} = \mathbf{r}. \end{cases} \tag{7.5}$$

Here, $\mathbf{c}^T$ is the transpose of the column vector $\mathbf{c}$. We use this notation later to simplify our analysis.

Example 7.3. Consider the problem of a toy company that produces toy planes and toy boats. The toy company can sell its planes for $10 and its boats for $8. It costs $3 in raw materials to make a plane and $2 in raw materials to make a boat. A plane requires 3 hours to make and 1 hour to finish, while a boat requires 1 hour to make and 2 hours to finish. The toy company knows that it will not sell any more than 35 planes per week. Further, given the number of workers, the company cannot spend any more than 160 hours per week finishing toys and 120 hours per week making toys. The company wishes to maximize the profit it makes by choosing how much of each toy to produce.

We can represent the profit maximization problem of the company as a linear programming problem. Let x_1 be the number of planes the company will produce, and let x_2 be the number of boats the company will produce. The profit for each plane is $10 - $3 = $7 per plane, and the profit for each boat is $8 - $2 = $6 per boat. Thus, the total profit the company will make is

$$z(x_1, x_2) = 7x_1 + 6x_2. \tag{7.6}$$

The company can spend no more than 120 hours per week making toys, and since a plane takes 3 hours to make and a boat takes 1 hour to make, we have

$$3x_1 + x_2 \le 120. \tag{7.7}$$

Likewise, the company can spend no more than 160 hours per week finishing toys, and since it takes 1 hour to finish a plane and 2 hours to finish a boat, we have

$$x_1 + 2x_2 \le 160. \tag{7.8}$$

Finally, we know that $x_1 \le 35$ since the company will make no more than 35 planes per week. Thus, the complete linear programming problem is given by

$$\begin{cases} \max\ z(x_1, x_2) = 7x_1 + 6x_2, \\ s.t.\ 3x_1 + x_2 \leq 120, \\ \quad\quad x_1 + 2x_2 \leq 160, \\ \quad\quad x_1 \leq 35, \\ \quad\quad x_1 \geq 0, \\ \quad\quad x_2 \geq 0. \end{cases} \tag{7.9}$$

Remark 7.4. In strict terms, the linear programming problem in Example 7.3 is not a true linear programming problem because we don't want to manufacture a fractional number of boats or planes; therefore, x_1 and x_2 must be drawn from the *integers* and not the real numbers (a requirement for a linear programming problem). This type of problem is called an integer programming problem [82]. In Example 7.3, we ignore this fact and assume that we can indeed manufacture a fractional number of boats and planes.

7.2   Intuition on the Solution of Linear Programs

Remark 7.5. Linear programs (LPs) with two variables can be solved graphically by plotting the feasible region along with the level curves of the objective function. We illustrate the method using the problem from Example 7.3.

Example 7.6 (Continuation of Example 7.3). To solve the linear programming problem from Example 7.3 graphically, begin by drawing the feasible region. This is shown in the blue shaded region of Fig. 7.1. The dashed lines are labeled by the corresponding constraints and make up the boundary of the feasible region, including the constraints $x_1, x_2 \geq 0$.

After plotting the feasible region, the next step is to plot the level curves of the objective function. In our problem, the level sets have the form

$$7x_1 + 6x_2 = c \implies x_2 = \frac{-7}{6}x_1 + \frac{c}{6}.$$

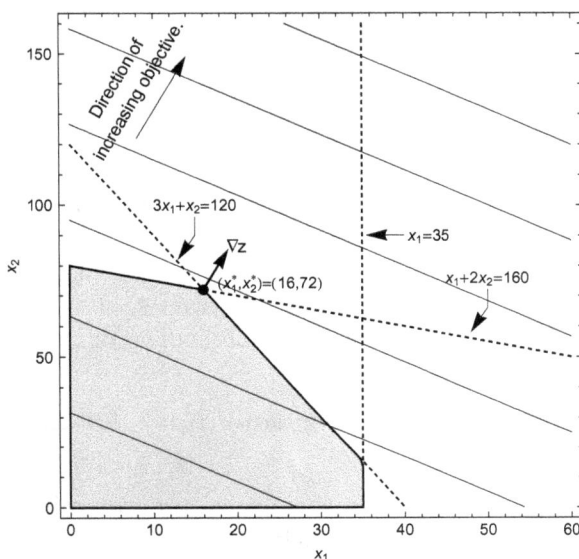

Fig. 7.1. The shaded region in the plot is the feasible region and represents the intersection of the five inequalities constraining the values of x_1 and x_2. The optimal solution is the "last" point in the feasible region that intersects a level set as we move in the direction of increasing profit.

This is a collection of parallel lines with slope $-\frac{7}{6}$ and intercept $\frac{c}{6}$, where c is varied as needed. In Fig. 7.1, the level curves are shown in colors ranging from purple to red depending upon the value of c. Larger values of c are closer to red.

To solve the linear programming problem, follow the level sets along the gradient (shown as the black arrow) until the last level set (line) intersects the feasible region. When doing this by hand, you can draw a single line of the form $7x_1 + 6x_2 = c$ and then simply draw parallel lines in the direction of the gradient $(7, 6)$. At some point, these lines will fail to intersect the feasible region. The last line to intersect the feasible region will do so at a point that maximizes the profit. In this case, the point that maximizes $z(x_1, x_2) = 7x_1 + 6x_2$, subject to the constraints given, is $(x_1^*, x_2^*) = (16, 72)$.

Note that the point of optimality $(x_1^*, x_2^*) = (16, 72)$ is at a corner of the feasible region. This corner is formed by the intersection of the two lines $3x_1 + x_2 = 120$ and $x_1 + 2x_2 = 160$. These lines correspond to the constraints

$$3x_1 + x_2 \leq 120,$$

$$x_1 + 2x_2 \leq 160.$$

At this point, both of these constraints are *binding*, while the other constraints are non-binding. (See Definition 6.11.) In general, when an optimal solution to a linear programming problem exists, it is at the intersection of several binding constraints; that is, it will occur at a vertex of a higher-dimensional polyhedron or polytope.

7.2.1 *Karush–Kuhn–Tucker conditions for linear programs*

Remark 7.7. As with any mathematical programming problem, we can derive the Karush–Kuhn–Tucker (KKT) conditions for a linear programming problem. We illustrate this by deriving the KKT conditions for Example 7.3. Note that, since linear (affine) functions are both convex and concave functions, we know that finding Lagrange multipliers satisfying the KKT conditions is necessary and sufficient for proving that a point is an optimal solution.

Example 7.8. Let $z(x_1, x_2) = 7x_1 + 6x_2$, the objective function in Eq. (7.9). We have argued that the point of optimality is $(x_1^*, x_2^*) = (16, 72)$. The KKT conditions for Eq. (7.9) are as follows:

Primal feasibility

$$
\begin{cases}
& \text{Lagrange Multiplier} \\
g_1(x_1^*, x_2^*) = 3x_1^* + x_2^* - 120 \leq 0 & (\lambda_1), \\
g_2(x_1^*, x_2^*) = x_1^* + 2x_2^* - 160 \leq 0 & (\lambda_2), \\
g_3(x_1^*, x_2^*) = x_1^* - 35 \leq 0 & (\lambda_3), \\
g_4(x_1^*, x_2^*) = -x_1^* \leq 0 & (\lambda_4), \\
g_5(x_1^*, x_2^*) = -x_2^* \leq 0 & (\lambda_5).
\end{cases}
\tag{7.10}
$$

Dual feasibility

$$\begin{cases} \nabla z(x_1^*, x_2^*) - \displaystyle\sum_{i=1}^{5} \lambda_i \nabla g_i(x_1^*, x_2^*) = \begin{bmatrix} 0 \\ 0 \end{bmatrix}, \\ \lambda_i \geq 0 \quad i = 1, \ldots, 5. \end{cases} \tag{7.11}$$

Complementary slackness

$$\{\lambda_i g_i(x_1^*, x_2^*) = 0 \quad i = 1, \ldots, 5. \tag{7.12}$$

The vector $\langle 0, 0 \rangle$ occurs in our dual feasible conditions because the gradients of our functions will all be two-dimensional vectors (there are two variables). Specifically, we can compute

$$\nabla z(x_1^*, x_2^*) = \langle 7, 6 \rangle,$$
$$\nabla g_1(x_1^*, x_2^*) = \langle 3, 1 \rangle,$$
$$\nabla g_2(x_1^*, x_2^*) = \langle 1, 2 \rangle,$$
$$\nabla g_3(x_1^*, x_2^*) = \langle 1, 0 \rangle,$$
$$\nabla g_4(x_1^*, x_2^*) = \langle -1, 0 \rangle,$$
$$\nabla g_5(x_1^*, x_2^*) = \langle 0, -1 \rangle.$$

Note that $g_3(16, 72) = 16 - 35 = -17 \neq 0$. This means that for complementary slackness to be satisfied, we must have $\lambda_2 = 0$. By the same reasoning, $\lambda_4 = 0$ because $g_4(16, 72) = -16 \neq 0$ and $\lambda_5 = 0$ because $g_5(16, 72) = -72 \neq 0$. Thus, dual feasibility can be simplified to

$$\begin{cases} \begin{bmatrix} 7 \\ 6 \end{bmatrix} - \lambda_1 \begin{bmatrix} 3 \\ 1 \end{bmatrix} - \lambda_2 \begin{bmatrix} 1 \\ 2 \end{bmatrix} = \begin{bmatrix} 0 \\ 0 \end{bmatrix} \\ \lambda_1, \lambda_2 \geq 0. \end{cases} \tag{7.13}$$

This is just a set of linear equations with some non-negativity constraints, which we ignore for now. We have

$$7 - 3\lambda_1 - \lambda_2 = 0 \implies 3\lambda_1 + \lambda_2 = 7, \tag{7.14}$$
$$6 - \lambda_1 - 2\lambda_2 = 0 \implies \lambda_1 + 2\lambda_2 = 6. \tag{7.15}$$

We can solve these linear equations (and hope that the solution is positive). Doing so yields

$$\lambda_1 = \frac{8}{5} \geq 0, \tag{7.16}$$

$$\lambda_2 = \frac{11}{5} \geq 0. \tag{7.17}$$

Thus, we have found a KKT point:

$$x_1^* = 16,$$
$$x_2^* = 72,$$
$$\lambda_1 = \frac{8}{5},$$
$$\lambda_2 = \frac{11}{5}, \tag{7.18}$$
$$\lambda_3 = 0,$$
$$\lambda_4 = 0,$$
$$\lambda_5 = 0.$$

This proves (via Theorem 6.36) that the point we found graphically is, in fact, the optimal solution to Eq. (7.9).

7.2.2 *Problems with an infinite number of solutions*

Remark 7.9. Linear programming problems can have an infinite number of solutions. We can construct such a problem by modifying the objective function in Example 7.3.

Example 7.10. Suppose the toy maker in Example 7.3 finds that he or she can sell planes for a profit of \$18 each instead of \$7 each. The new linear programming problem becomes

$$\begin{cases} \max \ z(x_1, x_2) = 18x_1 + 6x_2, \\ s.t. \ 3x_1 + x_2 \leq 120, \\ \quad x_1 + 2x_2 \leq 160, \\ \quad x_1 \leq 35, \\ \quad x_1 \geq 0, \\ \quad x_2 \geq 0. \end{cases} \tag{7.19}$$

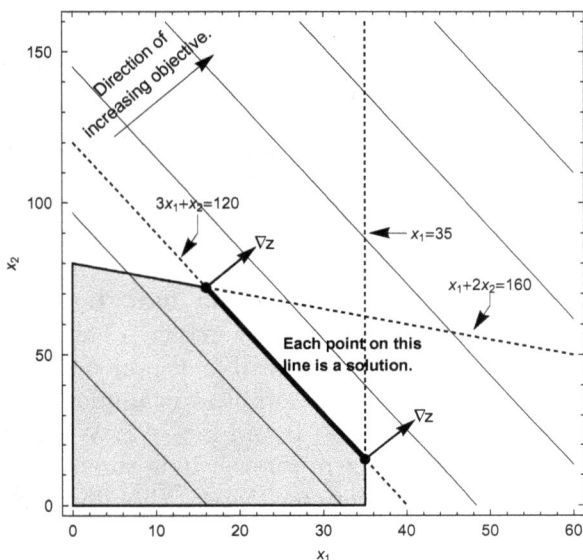

Fig. 7.2. An example of infinitely many alternative optimal solutions in a linear programming problem: The level curves for $z(x_1, x_2) = 18x_1 + 6x_2$ are *parallel* to one face of the polygonal boundary of the feasible region. Moreover, this side contains the points of greatest value for $z(x_1, x_2)$ inside the feasible region. Any combination of (x_1, x_2) on the line $3x_1 + x_2 = 120$ for $x_1 \in [16, 35]$ will provide the largest possible value $z(x_1, x_2)$ can take in the feasible region.

Applying the graphical method for finding optimal solutions to linear programming problems yields the plot shown in Fig. 7.2. The level curves for the function $z(x_1, x_2) = 18x_1 + 6x_2$ are *parallel* to one face of the polygonal boundary of the feasible region. In particular, they are parallel to the line defined by the equation $3x_1 + x_2 = 120$. Hence, as we move further up and to the right in the direction of the gradient, corresponding to larger and larger values of $z(x_1, x_2)$, we see that there is not one point on the boundary of the feasible region that intersects that level set with the greatest value, but instead a side of the polygonal boundary described by the line $3x_1 + x_2 = 120$, where $x_1 \in [16, 35]$ intersects the level set with the greatest value. Let

$$S = \{(x_1, x_2) | 3x_1 + x_2 \le 120, \ x_1 + 2x_2 \le 160, \ x_1 \le 35, \ x_1, x_2 \ge 0\},$$

that is, S is the feasible region of the problem. Then, for any value of $x_1^* \in [16, 35]$ and any value x_2^* so that $3x_1^* + x_2^* = 120$, we will have

$z(x_1^*, x_2^*) \geq z(x_1, x_2)$ for all $(x_1, x_2) \in S$. Since x_1 and x_2 may take on infinitely many values, we see that this problem has an infinite number of alternative optimal solutions.

Remark 7.11 (Other Possibilities). In addition to the two scenarios above, in which a linear programming problem has a unique solution or an infinite number of alternative optimal solutions, there are two other possibilities:

(1) A linear programming problem can have no solution, which occurs when the feasible region is empty. (The problem is said to be infeasible.) An empty feasible region occurs if there are inconsistent constraints. The simplest examples of such a situation are the constraints $x \geq 0$ and $x \leq -1$. When dealing with large linear programming problems, it is sometimes difficult to spot (visually) inconsistent constraints, though there are tests to determine whether constraints are inconsistent.
(2) The linear programming problem can have an unbounded solution, which can occur if the feasible region is an unbounded set and there are feasible solutions that make the objective function arbitrarily large in a maximization problem or arbitrarily small in a minimization problem.

Fortunately, we will not encounter either of those situations in our study of zero-sum games, and so we can ignore these possibilities. The interested reader can consult Ref. [78] or a related text on linear programming for details.

7.3 A Linear Program for Zero-Sum Game Players

Remark 7.12. We are now in a position to derive a linear programming problem for finding the Nash equilibrium of a zero-sum game.

Derivation 7.13. Let $\mathcal{G} = (\mathbf{P}, \Sigma, \mathbf{A})$ be a zero-sum game with $\mathbf{A} \in \mathbb{R}^{m \times n}$. We use the components of $\mathbf{A}$ and thus assume that

$$\mathbf{A} = \begin{bmatrix} a_{11} & a_{12} & \cdots & a_{1n} \\ a_{21} & a_{22} & \cdots & a_{2n} \\ \vdots & \vdots & \ddots & \vdots \\ a_{m1} & a_{m2} & \cdots & a_{mn} \end{bmatrix}.$$

Recall from Theorem 5.57 that the following are equivalent:

(1) There is a Nash equilibrium $(\mathbf{x}^*, \mathbf{y}^*)$ for $\mathcal{G}$.
(2) The following equation holds:

$$v_1 = \max_{\mathbf{x}} \min_{\mathbf{y}} \mathbf{x}^T \mathbf{A} \mathbf{y} = \min_{\mathbf{y}} \max_{\mathbf{x}} \mathbf{x}^T \mathbf{A} \mathbf{y} = v_2. \tag{7.20}$$

(3) There exists a real number v and $\mathbf{x}^* \in \Delta_m$ and $\mathbf{y}^* \in \Delta_n$ so that the following inequalities hold:

$$\sum_i \mathbf{A}_{ij} x_i^* \geq v, \quad \text{with } j \in \{1, \ldots, n\} \quad \text{and} \tag{7.21}$$

$$\sum_j \mathbf{A}_{ij} y_j^* \leq v, \quad \text{with } i \in 1, \ldots, m\}. \tag{7.22}$$

The fact that $\mathbf{x}^* \in \Delta_m$ implies that

$$x_1^* + \cdots + x_m^m = 1 \tag{7.23}$$

and $x*_i \geq 0$ for $i \in \{1, \ldots, m\}$. Similar conditions will hold for $\mathbf{y}^*$.

Combining Eq. (7.21) and incorporating the constraints imposed by $\mathbf{x}^* \in \Delta_m$, we arrive at a set of constraints for a linear programming problem. That is,

$$
\begin{aligned}
a_{11}x_1^* + \cdots + a_{m1}x_m^* - v &\geq 0, \\
a_{12}x_1^* + \cdots + a_{m2}x_m^* - v &\geq 0, \\
&\vdots \\
a_{1n}x_1^* + \cdots + a_{mn}x_m^* - v &\geq 0, \\
x_1^* + \cdots + x_m^* &= 1, \\
x_i^* \geq 0 \quad i \in \{1, \ldots, m\}.
\end{aligned}
\tag{7.24}
$$

In this set of constraints, we have $m + 1$ variables: $x_1^*, \ldots, x_m^*$ and v, the value of the game. We know that Player 1 (the row player) is a value maximizer. Therefore, Player 1 solves the linear programming problem

$$\begin{cases}
\max \ v, \\
s.t. \ a_{11}x_1^* + \cdots + a_{m1}x_m^* - v \geq 0, \\
\qquad a_{12}x_1^* + \cdots + a_{m2}x_m^* - v \geq 0, \\
\qquad \qquad \vdots \\
\qquad a_{1n}x_1^* + \cdots + a_{mn}x_m^* - v \geq 0, \\
\qquad x_1^* + \cdots + x_m^* = 1, \\
\qquad x_i^* \geq 0 \quad i \in \{1, \ldots, m\}.
\end{cases} \tag{7.25}$$

By a similar argument, we know that Player 2's equilibrium strategy $\mathbf{y}^*$ is constrained by

$$\begin{aligned}
a_{11}y_1^* + \cdots + a_{1n}y_n^* - v \leq 0, \\
a_{21}y_1^* + \cdots + a_{2n}y_n^* - v \leq 0, \\
\vdots \\
a_{m1}y_1^* + \cdots + a_{mn}y_n^* - v \leq 0, \\
y_1^* + \cdots + y_n^* = 1, \\
y_i^* \geq 0 \quad i \in \{1, \ldots, n\}.
\end{aligned} \tag{7.26}$$

We know that Player 2 (the column player) is a value minimizer. Therefore, Player 2 solves the linear programming problem

$$\begin{cases}
\min \ v, \\
s.t. \ a_{11}y_1 + \cdots + a_{1n}y_n - v \leq 0, \\
\qquad a_{21}y_1 + \cdots + a_{2n}y_n - v \leq 0, \\
\qquad \qquad \vdots \\
\qquad a_{m1}y_1 + \cdots + a_{mn}y_n - v \leq 0, \\
\qquad y_1 + \cdots + y_n = 1, \\
\qquad y_i \geq 0 \quad i \in \{1, \ldots, n\}.
\end{cases} \tag{7.27}$$

Thus, we have derived the two linear programming problems for players in a zero-sum game.

Example 7.14. Consider the game matrix from Example 5.3. In this example, two television networks (streaming services) were attempting to decide what content to provide. The payoff matrix for Player 1 is given as

$$\mathbf{A} = \begin{bmatrix} -15 & -35 & 10 \\ -5 & 8 & 0 \\ -12 & -36 & 20 \end{bmatrix}.$$

This is a zero-sum game, so the payoff matrix for Player 2 is simply the negation of this matrix. The linear programming problem for Player 1 is

$$\begin{aligned} \max \quad & v, \\ s.t. \quad & -15x_1 - 5x_2 - 12x_3 - v \geq 0, \\ & -35x_1 + 8x_2 - 36x_3 - v \geq 0, \\ & 10x_1 + 20x_3 - v \geq 0, \\ & x_1 + x_2 + x_3 = 1, \\ & x_1, x_2, x_3 \geq 0. \end{aligned} \tag{7.28}$$

Note that we simply work our way down each column of the matrix $\mathbf{A}$ to form the constraints of the linear programming problem. To form the problem for Player 2, we work our way across the rows of $\mathbf{A}$ and obtain

$$\begin{aligned} \min \quad & v, \\ s.t. \quad & -15y_1 - 35y_2 + 10y_3 - v \leq 0, \\ & -5y_1 + 8y_2 - v \leq 0, \\ & -12y_1 - 36y_2 + 20y_3 - v \leq 0, \\ & y_1 + y_2 + y_3 = 1, \\ & y_1, y_2, y_3 \geq 0. \end{aligned} \tag{7.29}$$

7.4 Solving Linear Programs Using a Computer

Remark 7.15. Solving linear programs can be accomplished by using the simplex algorithm or an interior point method, the details of which are outside the scope of this book. The interested reader should consult a text on linear programming, such as Ref. [78]. It suffices to say that each approach provides a series of steps that can be performed by a computer to solve a linear programming problem. While there are several software packages that can solve linear programming problems (e.g., Python's SciPy, MATLAB, etc.), we illustrate how to solve the problems in Eqs. (7.28) and (7.29) using Mathematica because it has a simple interface that most closely resembles the linear programming problems as given.

Example 7.16. Consider the linear programming problem in Eq. (7.28). A Mathematica code to solve this problem is shown in the following. Note that in Mathematica, comments are written as "(*Comment*)." These comments are optional, as is the spacing.

```
FindMaximum[
    {
        v,                              (*Objective*)
        -15*x1 - 5*x2 -12*x3 >= v,     (*Constraint 1*)
        -35*x1 + 8*x2 -36*x3 >= v,     (*Constraint 2*)
        10*x1          +20*x3 >= v,    (*Constraint 3*)
        x1 +    x2 +    x3 == 1,       (*Probability
            Constraint*)
        x1 >= 0, x2 >= 0, x3 >=0       (*Non-negativity
            Constraints*)
    },
    {x1, x2, x3, v} (*Variables*)
]
```

Note that we have replaced x_1, x_2, and x_3 with x1, x2, and x3, respectively, to ease the representation of the code in the text. The code makes use of **FindMaximum**, which can automatically deduce that a linear program is being passed into it. The answer reported by Mathematica is

```
{-5., {x1 -> 0., x2 -> 1., x3 -> 0., v -> -5.}}
```

This tells us that at equilibrium, Player 1 receives a payoff of -5 and should use a pure strategy, $\mathbf{e}_2 = \langle 0, 1, 0 \rangle$.

The corresponding code to solve Player 2's linear programming problem given in Eq. (7.29) is shown as follows.

```
FindMinimum[
    {
        v,                                          (*Objective*)
        -15*y1 - 35*y2 +  10*y3 <= v, (*Constraint 1*)
        -5*y1 +  8*y2           <= v, (*Constraint 2*)
        -12*y1 - 36*y2 + 20*y3 <= v, (*Constraint 3*)
        y1 +   y2 +     y3 == 1, (*Probability
            Constraint*)
        y1 >= 0, y2 >= 0, y3 >= 0    (*Non-negativity
            Constraints*)
    },
    {y1, y2, y3, v} (*Variables*)
]
```

This code makes use of the **FindMinimum** function, which operates in the same way as **FindMaximum** used above. The solution provided by Mathematica^{TM} is

$$\{-5., \{y1 \rightarrow 1., y2 \rightarrow 0., y3 \rightarrow 0., v \rightarrow -5.\}\}$$

This tells us that at equilibrium, Player 2 should use a pure strategy, $\mathbf{e}_1 = \langle 1, 0, 0 \rangle$. The payoff at equilibrium to Player 2 is not -5; this occurs because we are using Player 1's payoff matrix. Instead, the payoff to Player 2 at equilibrium is $-(-5) = 5$.

Thus, we have used two linear programming problems to confirm that $(\mathbf{e}_2, \mathbf{e}_1)$ is the Nash equilibrium game between two networks from Example 5.3, and the value of the game is $v = -5 = \mathbf{e}_2^T \mathbf{A} \mathbf{e}_1 = A_{21}$. This is precisely the answer we got using the minimax theorem.

7.5   Standard Form, Slack and Surplus Variables

Remark 7.17. Before completing our analysis of zero-sum games and linear programming, we require one additional definition for completeness.

Definition 7.18 (Standard Form). A linear programming problem is in *standard form* if it is written with the constraints $\mathbf{Ax} = \mathbf{b}$

and $\mathbf{x} \geq \mathbf{0}$, as in

$$
\begin{cases}
\max & z(\mathbf{x}) = \mathbf{c}^T\mathbf{x}, \\
s.t. & \mathbf{Ax} = \mathbf{b}, \\
& \mathbf{x} \geq \mathbf{0}.
\end{cases}
\tag{7.30}
$$

We note that this could be a minimization problem as well. It is the structure of the constraints that is important.

Remark 7.19. It is relatively easy to convert any inequality constraint into an equality constraint. Consider the inequality constraint

$$
a_{i1}x_1 + a_{i2}x_2 + \cdots + a_{in}x_n \leq b_i.
\tag{7.31}
$$

We can add a new *slack variable*, $s_i \geq 0$, to this constraint to obtain

$$
a_{i1}x_1 + a_{i2}x_2 + \cdots + a_{in}x_n + s_i = b_i.
$$

The slack variable then becomes just another variable whose value we must discover as we solve the linear program for which Eq. (7.31) is a constraint, as is the non-negativity of s_i. That is, we add $s_i \geq 0$ as a constraint to the modified linear programming problem as well.

We can deal with constraints of the form

$$
a_{i1}x_1 + a_{i2}x_2 + \cdots + a_{in}x_n \geq b_i
\tag{7.32}
$$

similarly. In this case, we subtract a *surplus variable*, $s_i \geq 0$, to obtain

$$
a_{i1}x_1 + a_{i2}x_2 + \cdots + a_{in}x_n - s_i = b_i.
$$

Thus, we have shown that any linear programming problem can be converted into a problem in standard form.

Example 7.20. We can convert Player 1's problem from Eq. (7.28) into the standard form by adding three surplus variables, s_1, s_2, and s_3. The resulting problem is

$$
\begin{aligned}
\max \quad & v \\
s.t. \quad & -15x_1 - 5x_2 - 12x_3 - v - s_1 = 0, \\
& -35x_1 + 8x_2 - 36x_3 - v - s_2 = 0, \\
& 10x_1 + 20x_3 - v - s_3 = 0, \\
& x_1 + x_2 + x_3 = 1, \\
& x_1, x_2, x_3, s_1, s_2, s_3 \geq 0.
\end{aligned}
\tag{7.33}
$$

7.6 Optimality Conditions for Zero-Sum Games and Duality

Remark 7.21. Our final goal in this chapter is to prove a deep connection between the linear programming problems for the two players. To achieve this, we make use of the concept of *duality*, which we define as we proceed. This section is dense, but the reward for working through it is an interesting theoretical property of zero-sum games. Perhaps more importantly, we will use this same approach when we tackle general-sum games in the following chapter.

Theorem 7.22. *Let* $\mathcal{G} = (\mathbf{P}, \Sigma, \mathbf{A})$ *be a zero-sum two-player game with* $\mathbf{A} \in \mathbb{R}^{m \times n}$. *Then, the linear program for Player 1,*

$$\max \ v$$
$$s.t. \ a_{11}x_1 + \cdots + a_{m1}x_m - v \geq 0,$$
$$a_{12}x_1 + \cdots + a_{m2}x_m - v \geq 0,$$
$$\vdots$$
$$a_{1n}x_1 + \cdots + a_{mn}x_m - v \geq 0,$$
$$x_1 + \cdots + x_m - 1 = 0,$$
$$x_i \geq 0 \quad i = 1, \ldots, m,$$

has an optimal solution, $\mathbf{x} = \langle x_1, \ldots, x_m \rangle$, *if and only if there exist Lagrange multipliers,* $y_1, \ldots, y_n, \rho_1, \ldots, \rho_m$, *and* ν, *and surplus variables,* $s_1, \ldots, s_n$, *such that the following hold:*

$$\text{Primal feasibility}: \begin{cases} \displaystyle\sum_{i=1}^{m} a_{ij}x_i - v - s_j = 0 & j \in \{1, \ldots, n\}, \\[2mm] \displaystyle\sum_{i=1}^{m} x_i = 1, \\[2mm] x_i \geq 0 & i \in \{1, \ldots, m\}, \\[1mm] s_j \geq 0 & j \in 1, \ldots, n\}, \\[1mm] v \quad \text{unrestricted}. \end{cases}$$

$$Dual\ feasibility : \begin{cases} \displaystyle\sum_{j=1}^{n} a_{ij}y_j - \nu + \rho_i = 0 & i \in \{1,\ldots,m\}, \\[2em] \displaystyle\sum_{j=1}^{n} y_j = 1, \\[1em] y_j \geq 0 & j \in \{1,\ldots,n\}, \\[0.5em] \rho_i \geq 0 & i \in \{1,\ldots,m\}, \\[0.5em] \nu \quad unrestricted. \end{cases}$$

$$Complementary\ slackness : \begin{cases} y_j s_j = 0 & j \in \{1,\ldots,n\}, \\ \rho_i x_i = 0 & i \in \{1,\ldots,m\}. \end{cases}$$

Proof.   We begin by showing the statements that primal feasibility must hold. Clearly, v is unrestricted and $x_i \geq 0$ for $i \in \{1,\ldots,m\}$. The fact that $x_1 + \cdots + x_m = 1$ is also clear from the statement of the problem. We can rewrite each constraint of the form

$$a_{1j}x_1 + \cdots + a_{mj}x_m - v \geq 0, \tag{7.34}$$

where $j \in \{1,\ldots,n\}$, as

$$a_{1j}x_1 + \cdots + a_{mj}x_m - v - s_j = 0, \tag{7.35}$$

where $s_j \geq 0$ is a surplus variable for $j \in \{1,\ldots,n\}$. From this, it is clear that if $\langle x_1,\ldots,x_m\rangle$ is a feasible solution, then at least the variables $s_1,\ldots,s_n \geq 0$ exist and primal feasibility must hold.

We now prove that complementary slackness and dual feasibility hold. Rewrite the constraints of the form in Eq. (7.34) as

$$-a_{1j}x_1 - \cdots - a_{mj}x_m + v \leq 0 \quad j \in \{1,\ldots,n\} \tag{7.36}$$

and each non-negativity constraint as

$$-x_i \leq 0 \quad i \in \{1,\ldots,m\}. \tag{7.37}$$

We know that each affine function is both concave and convex; therefore, by Theorem 6.36 (the KKT theorem), there are Lagrange multipliers, $y_1,\ldots,y_n$, corresponding to the constraints in Eq. (7.36) and

Lagrange multipliers, $\rho_1, \ldots, \rho_m$, corresponding to the constraints in Eq. (7.37). Lastly, there is a Lagrange multiplier, ν, corresponding to the constraint

$$x_1 + x_2 + \cdots + x_m - 1 = 0. \tag{7.38}$$

From Theorem 6.36, we know that

$$y_j \geq 0 \quad j \in \{1, \ldots, n\},$$
$$\rho_i \geq 0 \quad i \in \{1, \ldots, m\},$$
$$\nu \quad \text{unrestricted.}$$

To see that complementary slackness holds, note that by Theorem 6.36, we know that

$$y_j \left(-a_{1j}x_1 - \cdots - a_{mj}x_m + \nu\right) = 0 \quad j \in \{1, \ldots, n\},$$
$$\rho_i(-x_i) = 0 \quad i \in \{1, \ldots, m\}.$$

If $\rho_i(-x_i) = 0$, then $\rho_i x_i = 0$ for all i. From Eq. (7.35),

$$a_{1j}x_1 + \cdots + a_{mj}x_m - \nu + s_j = 0 \implies s_j = -a_{1j}x_1 - \cdots - a_{mj}x_m + \nu.$$

Therefore, we can write

$$y_j \left(-a_{1j}x_1 - \cdots - a_{mj}x_m + \nu\right) = 0 \implies y_j(s_j) = 0,$$

for all j. Thus, we have shown that

$$y_j s_j = 0 \quad j \in \{1, \ldots, n\}, \tag{7.39}$$
$$\rho_i x_i = 0 \quad i \in \{1, \ldots, m\}. \tag{7.40}$$

We now complete the proof by showing that dual feasibility holds. Let

$$g_j(x_1, \ldots, x_m, \nu) = -a_{1j}x_1 - \cdots - a_{mj}x_m + \nu \quad j \in \{1, \ldots, n\},$$
$$f_i(x_1, \ldots, x_m, \nu) = -x_i \quad i \in \{1, \ldots, m\},$$
$$h(x_1, \ldots, x_m, \nu) = x_1 + x_2 + \cdots + x_m - 1,$$
$$z(x_1, \ldots, x_m, \nu) = \nu.$$

Then, we can apply Theorem 6.36 to see that

$$\nabla z - \sum_{j=1}^{n} y_j \nabla g_j(x_1, \ldots, x_n, n)$$

$$- \sum_{i=1}^{m} \rho_i \nabla f_i(x_1, \ldots, x_m, v) - \nu \nabla h(x_1, \ldots, x_m, v) = 0. \qquad (7.41)$$

Computing the gradients yields

$$\nabla z(x_1, \ldots, x_m, v) = \begin{bmatrix} 0 \\ 0 \\ \vdots \\ 0 \\ 1 \end{bmatrix} \in \mathbb{R}^{(m+1) \times 1},$$

$$\nabla h(x_1, \ldots, x_m, v) = \begin{bmatrix} 1 \\ 1 \\ \vdots \\ 1 \\ 0 \end{bmatrix} \in \mathbb{R}^{(m+1) \times 1},$$

$$\nabla f_i(x_1, \ldots, x_m, v) = -\mathbf{e}_i \in \mathbb{R}^{(m+1) \times 1},$$

and

$$\nabla g_j(x_1, \ldots, x_m, v) = \begin{bmatrix} -a_{1j} \\ -a_{2j} \\ \vdots \\ -a_{mj} \\ 1 \end{bmatrix} \in \mathbb{R}^{(m+1) \times 1}.$$

Before proceeding, note that in computing $\nabla f_i(x_1, \ldots, x_m, v) = -\mathbf{e}_i \in \mathbb{R}^{(m+1) \times 1}$ with $i \in \{1, \ldots, m\}$, we will *never* see the vector

$$-\mathbf{e}_{m+1} = \begin{bmatrix} 0 \\ 0 \\ \vdots \\ 0 \\ -1 \end{bmatrix} \in \mathbb{R}^{(m+1) \times 1}$$

because there is *no* function $f_{m+1}(x_1, \ldots, x_m, v)$. We can now rewrite Eq. (7.41) as

$$\begin{bmatrix} 0 \\ 0 \\ \vdots \\ 0 \\ 1 \end{bmatrix} - \left(\sum_{j=1}^{n} y_j \begin{bmatrix} -a_{1j} \\ -a_{2j} \\ \vdots \\ -a_{mj} \\ 1 \end{bmatrix} \right) - \left(\sum_{i=1}^{m} \rho_i(-\mathbf{e}_i) \right) - \nu \begin{bmatrix} 1 \\ 1 \\ \vdots \\ 1 \\ 0 \end{bmatrix} = \mathbf{0}. \quad (7.42)$$

Consider the ith row of that expression, where $1 \le i \le m$. Adding term-by-term, we have

$$0 + \sum_{j=1}^{n} a_{ij} y_j + \rho_i - \nu = 0. \quad (7.43)$$

Now, consider row $m + 1$. We have

$$1 - \sum_{j=1}^{n} y_j + 0 + 0 = 0. \quad (7.44)$$

From these two equations, we conclude that

$$\sum_{j=1}^{n} \mathbf{A}_{ij} y_j + \rho_i - \nu = 0,$$

$$\sum_{j=1}^{n} y_j = 1.$$

Thus, we have shown that dual feasibility holds. The fact that these conditions are necessary and sufficient follows from the fact that a linear programming problem is a convex optimization problem and from Theorem 6.36. $\square$

Remark 7.23. The proof of the following theorem is nearly identical to the proof of the previous theorem. It is left to the reader as an exercise.

Theorem 7.24. *Let $\mathcal{G} = (\mathbf{P}, \Sigma, \mathbf{A})$ be a zero-sum two-player game with $\mathbf{A} \in \mathbb{R}^{m \times n}$. Then, the linear program for Player 2,*

$$\min \ \nu$$

$$\text{s.t.} \ a_{11}y_1 + \cdots + a_{1n}y_n - \nu \le 0,$$

$$a_{21}y_1 + \cdots + a_{2n}y_n - \nu \le 0,$$

$$\vdots$$

$$a_{m1}y_1 + \cdots + a_{mn}y_n - \nu \le 0,$$

$$y_1 + \cdots + y_n - 1 = 0,$$

$$y_i \ge 0 \quad i = 1, \ldots, m,$$

has an optimal solution, $\mathbf{y} = \langle y_1, \ldots, y_n \rangle$, if and only if there exist Lagrange multipliers, $x_1, \ldots, x_m$, $s_1, \ldots, s_n$, and v, and slack variables, $\rho_1, \ldots, \rho_m$, such that we have the following:

Primal feasibility :
$$\begin{cases} \sum_{j=1}^{n} a_{ij} y_j - \nu + \rho_i = 0 \quad i \in \{1, \ldots, m\}, \\[2mm] \sum_{j=1}^{n} y_j = 1, \\[2mm] y_j \ge 0 \quad j \in \{1, \ldots, n\}, \\[1mm] \rho_i \ge 0 \quad i \in \{1, \ldots, m\}, \\[1mm] \nu \quad \text{unrestricted.} \end{cases}$$

Dual feasibility :
$$\begin{cases} \sum_{i=1}^{m} a_{ij} x_i - v - s_j = 0 \quad j \in \{1, \ldots, n\}, \\[2mm] \sum_{i=1}^{m} x_i = 1, \\[2mm] x_i \ge 0 \quad i \in \{1, \ldots, m\}, \\[1mm] s_j \ge 0 \quad j \in \{1, \ldots, n\}, \\[1mm] v \quad \text{unrestricted.} \end{cases}$$

$$\text{Complementary slackness}: \begin{cases} y_j s_j = 0 & j \in \{1, \ldots, n\}, \\ \rho_i x_i = 0 & i \in \{1, \ldots, m\}. \end{cases}$$

Remark 7.25. Theorems 7.22 and 7.24 show that the KKT conditions for the linear programming problems for Player 1 and Player 2 in a zero-sum game are identical, but with the primal and dual feasibility conditions exchanged. Linear programming problems with this property are naturally related, as we discuss in the following.

Definition 7.26. Let P and D be two linear programming problems. If the KKT conditions for Problem P are equivalent to the KKT conditions for Problem D with primal feasibility and dual feasibility exchanged, then Problems P and D are called *dual linear programming problems*.

Corollary 7.27. *The linear programming problem for Player 1 is the dual problem of the linear programming problem for Player 2 in a zero-sum two-player game, $\mathcal{G} = (\mathbf{P}, \Sigma, \mathbf{A})$, with $\mathbf{A} \in \mathbb{R}^{m \times n}$.*

Remark 7.28. There is a very deep theorem about dual linear programming problems, which is beyond the scope of this book. The interested reader can consult Ref. [78] for a proof. We make use of it to prove the minimax theorem in a completely novel way.

Theorem 7.29 (Strong Duality Theorem). *Let P and D be dual linear programming problems. Then, exactly one of the following statements holds:*

(1) *Both P and D have a solution, and at optimality, the objective function value for Problem P is identical to the objective function value for Problem D.*
(2) *Problem P has no solution because it is unbounded, and Problem D has no solution because it is infeasible.*
(3) *Problem D has no solution because it is unbounded, and Problem P has no solution because it is infeasible.*
(4) *Both Problem P and Problem D are infeasible.*

Theorem 7.30 (Minimax Theorem). *Let $\mathcal{G} = (\mathbf{P}, \Sigma, \mathbf{A})$ be a zero-sum two-player game with $\mathbf{A} \in \mathbb{R}^{m \times n}$, then there exists a Nash equilibrium, $(\mathbf{x}^*, \mathbf{y}^*) \in \Delta$. Furthermore, for every Nash equilibrium strategy pair, $(\mathbf{x}^*, \mathbf{y}^*) \in \Delta$, there is one unique value, $v^* = \mathbf{x}^{*T} \mathbf{A} \mathbf{y}^*$.*

Sketch of Proof. Let Problems P_1 and P_2 be the linear programming problems in Eqs. (7.25) and (7.27) for Players 1 and 2, respectively. We have already shown these linear programming problems are dual; therefore, if Problem P_1 has a solution, then so does problem P_2. More importantly, at these optimal solutions $(\mathbf{x}^*, v^*), (\mathbf{y}^*, \nu^*)$, we know that $v^* = \nu^*$ as the objective function values must be equal by Theorem 7.29.

Consider Problem P_1: We know that $\mathbf{x} = \langle x_1, \ldots, x_m \rangle \in \Delta_m$ and Δ_m must be bounded. The value v clearly cannot exceed $\max_{ij} \mathbf{A}_{ij}$ as a result of the constraints and the fact that $\mathbf{x} \in \Delta_m$. Obviously, v can be made as small as we like; however, this is not possible since this is a maximization problem. The fact that v is bounded from above and $\mathbf{x} \in \Delta_m$ and P_1 is a maximization problem (on v) implies that there is at least one solution $(\mathbf{x}^*, v^*)$ to Problem P_1. Likewise, there is a solution (y^*, ν^*) to Problem P_2 and $v^* = \nu^*$. Since the constraints for Problems P_1 and P_2 were taken from Theorem 5.57, we know that $(\mathbf{x}^*, \mathbf{y}^*)$ is a Nash equilibrium, and therefore such an equilibrium must exist.

Furthermore, while we have not proved this explicitly, one can prove that if $(\mathbf{x}^*, \mathbf{y}^*)$ is a Nash equilibrium, then it must be a part of a set of solutions $(\mathbf{x}^*, v^*), (\mathbf{y}^*, \nu^*)$ to Problems P_1 and P_2. Thus, any two equilibrium solutions are simply *alternative optimal solutions* to P_1 and P_2, respectively. Thus, for any Nash equilibrium pair, we have

$$\nu^* = v^* = \mathbf{x}^{*T} \mathbf{A} \mathbf{y}^*. \tag{7.45}$$

This completes the proof sketch. □

Remark 7.31 (A Remark on Complementary Slackness). Consider the KKT conditions for Players 1 and 2 given in Theorems 7.22 and 7.24. Suppose that in an optimal solution of the problem for Player 1, $s_j > 0$. Then, it follows that $y_j = 0$ by complementary slackness. We can understand this from a game-theoretic perspective. The expression

$$a_{1j} x_1 + \cdots + a_{mj} x_m$$

is the expected payoff to Player 1, assuming Player 2 plays column j. If $s_j > 0$, then

$$a_{1j} x_1 + \cdots + a_{mj} x_m > v.$$

However, that means if Player 2 *ever* played column j, then Player 1 could do better than the equilibrium value v of the game. Thus, Player 2 has no incentive to ever play this strategy, and the result is that $y_j = 0$ (as required by complementary slackness).

This implies something stronger. Let $(\mathbf{x}^*, \mathbf{y}^*)$ be a Nash equilibrium strategy pair. If $y_j^* > 0$, then

$$a_{1j}x_1^* + \cdots + a_{mj}x_m^* = v.$$

Thus, if Player 2 were to play the pure strategy $\mathbf{e}_j$ against $\mathbf{x}^*$, he would expect to receive a payoff of $-v$, which is precisely what he would receive using $\mathbf{y}^*$. That is, if $y_j^* > 0$, then

$$\mathbf{x}^{*T}\mathbf{A}\mathbf{y}^* = \mathbf{x}^{*T}\mathbf{A}\mathbf{e}_j.$$

A similar statement is true for Player 1, and thus we see that complementary slackness also implies the indifference theorem, Theorem 5.46.

7.7 Chapter Notes

The problem of solving a system of linear inequalities was studied as early as the 1800s, and a solution was given by Fourier. This method is now called the Fourier–Motzkin [83] method and is not usually covered in modern treatments of linear programming (except in its historical context). Foundational contributions to linear programming were made by George B. Dantzig, whose solution to two open problems in statistics (which he mistook for homework as a student) [84–86] may have formed the basis for part of the script of *Good Will Hunting*. Dantzig developed the simplex algorithm for solving arbitrary linear programming problems and is considered the father of modern algorithmic optimization. He first proved the duality results discussed in this chapter; however, it was von Neumann who conjectured them on meeting with Dantzig to discuss linear programming [87]. Dantzig's work (at first derided by colleagues for being linear in a nonlinear world) was defended by von Neumann. Interestingly, linear programming problems are often at the core of many problems in optimization [85, 86].

The connection between zero-sum games and linear programming is substantially deeper than the material in this chapter suggests.

We have proved that finding a Nash equilibrium strategy for a zero-sum game is equivalent to solving a pair of dual linear programming problems. Dantzig asserted that every linear programming problem (and its dual) can be converted into a corresponding zero-sum game [88]. His proof is incomplete and, as a result, so are the proofs presented in several textbooks, including Luce and Raiffa [1] and Raghavan [89]. A complete proof was only given by Adler in 2013 [90]. Thus, the study of zero-sum games is essentially the study of linear programs.

Using the equivalence, we can deduce how hard it is to solve a linear programming problem. Solving a linear programming problem can be accomplished in polynomial time [91]. By this, we mean that a solution can be produced using a set of steps, the number of which is a polynomial function of the size of the problem, as measured in the number of constraints and variables. This relatively recent result was improved upon and made practical by Karmarker's interior-point method [92], which has spawned an entirely new field in computational optimization. As a consequence, we can conclude that zero-sum games are relatively "easy" to solve. This is not true of general-sum games, which we discuss in the following chapter.

$$- \spadesuit \clubsuit \heartsuit \diamondsuit -$$

7.8   Exercises

7.1   A chemical manufacturer produces three chemicals: A, B, and C. These chemicals are produced by two processes: 1 and 2. Running process 1 for 1 hour costs \$4 and yields 3 units of chemical A, 1 unit of chemical B, and 1 unit of chemical C. Running process 2 for 1 hour costs \$1 and produces 1 unit of chemical A and 1 unit of chemical B (but none of chemical C). To meet customer demand, at least 10 units of chemical A, 5 units of chemical B, and 3 units of chemical C must be produced daily. Assume that the chemical manufacturer wants to minimize the cost of production. Develop a linear programming problem describing the constraints and objectives of the chemical manufacturer. [Hint: Let x_1 be the amount of time Process 1 is executed, and let x_2 be the amount of time Process 2 is executed. Use the coefficients above to express the cost of running Process 1 for x_1 time and Process 2 for x_2 time. Do the same to compute the amounts of chemicals A, B, and C that are produced.]

7.2 Modify the linear programming problem from Exercise 7.1 to obtain a linear programming problem with an infinite number of alternative optimal solutions. Solve the new problem and obtain a description for the set of alternative optimal solutions.

7.3 Construct the two linear programming problems for Bradley and von Kluge in the Battle of Avranches.

7.4 Prove Theorem 7.24.

7.5 Show that the Nash equilibrium given in Example 5.65 solves the linear programming problems from Exercise 7.3.

7.6 Use the approach in Remark 7.31 to prove the indifference theorem for Player 1.

Chapter 8

Quadratic Programs and General-Sum Games

Chapter Goals: In this chapter, we show how to convert any two-player bimatrix game into an optimization problem that has a quadratic objective function and linear constraints. Such a problem is called a *quadratic programming problem*. Readily available computer software can then be used to find the Nash equilibria as optimal solutions to these optimization problems. When combined with the results in Chapter 7, this provides an algorithmic way to solve any two-player (bi)matrix game. We achieve these results by exploring the Karush–Kuhn–Tucker (KKT) conditions of the individual player's optimization problems, showing how to combine them into a single quadratic problem whose solution yields the Nash equilibrium.

8.1 Introduction to Quadratic Programming

Definition 8.1 (Quadratic Programming Problem). Let $\mathbf{Q} \in \mathbb{R}^{n \times n}$ be a symmetric matrix, and let $\mathbf{c} \in \mathbb{R}^{n \times 1}$ be a vector. As in the definition of the general linear programming problem in Eq. (7.5), let matrix $\mathbf{A} \in \mathbb{R}^{m \times n}$, vector $\mathbf{b} \in \mathbb{R}^{m \times 1}$, matrix $\mathbf{H} \in \mathbb{R}^{l \times n}$, and vector $\mathbf{r} \in \mathbb{R}^{l \times 1}$. Then, a quadratic (maximization) programming problem is the nonlinear programming problem

$$QP \begin{cases} \max \ \dfrac{1}{2}\mathbf{x}^T \mathbf{Q}\mathbf{x} + \mathbf{c}^T \mathbf{x}, \\ s.t. \ \mathbf{A}\mathbf{x} \le \mathbf{b}, \\ \mathbf{H}\mathbf{x} = \mathbf{r}. \end{cases} \tag{8.1}$$

Example 8.2. Example 6.1 is an instance of a quadratic programming problem. Recall that we had

$$\begin{cases} \max \ A(x, y) = xy, \\ s.t. \ 2x + 2y = 100, \\ x \ge 0, \\ y \ge 0. \end{cases}$$

We can write this as

$$\begin{cases} \max \ \dfrac{1}{2}[x \ y] \begin{bmatrix} 0 & 1 \\ 1 & 0 \end{bmatrix} \begin{bmatrix} x \\ y \end{bmatrix}, \\ s.t. \ [2 \ 2] \begin{bmatrix} x \\ y \end{bmatrix} = 100, \\ \begin{bmatrix} x \\ y \end{bmatrix} \ge \begin{bmatrix} 0 \\ 0 \end{bmatrix}. \end{cases}$$

Obviously, we can put this problem in precisely the format given in Eq. (8.1), if so desired.

Remark 8.3. Quadratic programs are simply a special instance of nonlinear (or mathematical) programming problems. There are many applications for quadratic programs that are beyond the scope of this book, as well as many solution techniques for quadratic programs. Interested readers should consult Refs. [69, 70, 81] for details.

8.2   Solving Quadratic Programming Problems Using Computers

Remark 8.4. Just as was the case for linear programming problems, there are several computer-based solvers for quadratic programming problems. We illustrate how to use Mathematica to solve one such problem.

Example 8.5. We use Mathematica to solve the goat pen problem from Examples 6.1 and 8.2. The code to do this is shown as follows.

```
NMaximize[
    {
        x*y,                  (*Objective *)
        2*x + 2*y == 100,  (*Constraint*)
        x>=0, y>=0         (*Non-negativity
            Constraints*)
    },
    {x,y} (*Variables*),
    Method -> "DifferentialEvolution" (*Method to use*)
]
```

This code uses the **NMaximize** function, which will look for global optimal solutions numerically as opposed to local optimal solutions. This function can be customized by choosing a different method. We will need global optimal solutions when seeking Nash equilibria with quadratic programs. The output is

$$\{625., \{x \to 25., y \to 25.\}\}$$

Remark 8.6. Note that the input syntax for Mathematica is relatively natural. Other solvers such as Gurobi [93] or CPLEX [94] have a less natural syntax, but are generally more powerful. When using a solver, care must be taken. Many quadratic programming solvers require the problem to be convex. That is, they require the objective function,

$$z(\mathbf{x}) = \frac{1}{2}\mathbf{x}^T \mathbf{Q}\mathbf{x} + \mathbf{c}^T \mathbf{x},$$

to be convex. The quadratic programming problem we construct for finding Nash equilibria will not have this property as a general rule.

8.3 General-Sum Games and Quadratic Programming

Remark 8.7. The material in this section can be found originally in the work of Mangaserian and Stone [95], which we discuss in the chapter note.

Derivation 8.8. Consider a two-player general-sum game $\mathcal{G} = (\mathbf{P}, \Sigma, \mathbf{A}, \mathbf{B})$, with $\mathbf{A}, \mathbf{B} \in \mathbb{R}^{m \times n}$. Let $\mathbf{1}_m \in \mathbb{R}^{m \times 1}$ be the vector of all ones with m elements, and let $\mathbf{1}_n \in \mathbb{R}^{n \times 1}$ be the vector of all ones with n elements. By Theorem 5.75, there is at least one Nash equilibrium $(\mathbf{x}^*, \mathbf{y}^*)$. Suppose we treat this information as exogenous, i.e., provided somehow through a mechanism outside the problem. Then, the optimization problems for the two players are

$$P_1 \begin{cases} \max \ \mathbf{x}^T \mathbf{A} \mathbf{y}^*, \\ s.t. \ \mathbf{1}_m^T \mathbf{x} = 1, \\ \mathbf{x} \geq \mathbf{0}; \end{cases}$$

$$P_2 \begin{cases} \max \ \mathbf{x}^{*T} \mathbf{B} \mathbf{y}, \\ s.t. \ \mathbf{1}_n^T \mathbf{y} = 1, \\ \mathbf{y} \geq \mathbf{0}. \end{cases}$$

Individually, these are linear programs. Unfortunately, the Nash equilibrium $(\mathbf{x}^*, \mathbf{y}^*)$ is not available *a priori*. However, we can draw insight from these problems to derive the necessary and sufficient conditions for a Nash equilibrium.

Remark 8.9. In the proof of the following lemma, we use ∇ in multiple contexts, and so we append a subscript to it to indicate the variables involved in the differentiation. For example, if $\mathbf{x} = \langle x_1, \ldots, x_m \rangle$, then

$$\nabla_{\mathbf{x}} = \left\langle \frac{\partial}{\partial x_1}, \ldots, \frac{\partial}{\partial x_m} \right\rangle.$$

Likewise, if $\mathbf{y} = \langle y_1, \ldots, y_n \rangle$, then

$$\nabla_{\mathbf{y}} = \left\langle \frac{\partial}{\partial y_1}, \ldots, \frac{\partial}{\partial y_n} \right\rangle.$$

Remark 8.10. The proofs of the following lemma and theorem are long. However, they make up the core of the basic theory surrounding general-sum games. On a first reading, it is perfectly fine to skip the details of the proofs and move on to the example. However, it is worth the trouble to understand the proofs because they elucidate the connections between the KKT conditions and the Nash equilibria in general-sum games.

Lemma 8.11. *Let* $\mathcal{G} = (\mathbf{P}, \Sigma, \mathbf{A}, \mathbf{B})$ *be a general sum, two-player bimatrix game, with matrices* $\mathbf{A}, \mathbf{B} \in \mathbb{R}^{m \times n}$. *A pair* $(\mathbf{x}^*, \mathbf{y}^*) \in \Delta$ *is a Nash equilibrium if and only if there exist real values* α *and* β *such that*

$$\mathbf{x}^{*T} \mathbf{A} \mathbf{y}^* - \alpha = 0,$$

$$\mathbf{x}^{*T} \mathbf{B} \mathbf{y}^* - \beta = 0,$$

$$\mathbf{A} \mathbf{y}^* - \alpha \mathbf{1}_m \leq \mathbf{0},$$

$$\mathbf{x}^{*T} \mathbf{B} - \beta \mathbf{1}_n^T \leq \mathbf{0},$$

$$\mathbf{1}_m^T \mathbf{x}^* - 1 = 0,$$

$$\mathbf{1}_n^T \mathbf{y}^* - 1 = 0,$$

$$\mathbf{x}^* \geq \mathbf{0},$$

$$\mathbf{y}^* \geq \mathbf{0}.$$

Proof. Assume that $\mathbf{x}^* = \langle x_1^*, \ldots, x_m^* \rangle$ and $\mathbf{y}^* = \langle y_1^*, \ldots, y_n^* \rangle$. Consider the KKT conditions for the linear programming problem for P_1. The objective function is

$$z(x_1, \ldots, x_n) = \mathbf{x}^T \mathbf{A} \mathbf{y}^* = \mathbf{c}^T \mathbf{x},$$

where $\mathbf{c} \in \mathbb{R}^{n \times 1}$ and

$$c_i = \mathbf{A}_{i \cdot} \mathbf{y}^* = a_{i1} y_1^* + a_{i2} y_2^* + \cdots + a_{in} y_n^*.$$

The vector $\mathbf{x}^*$ is an optimal solution for this problem if and only if there are Lagrange multipliers $\lambda_1, \ldots, \lambda_m$, corresponding to the constraints $\mathbf{x} \geq \mathbf{0}$, and α, corresponding to the constraint $\mathbf{1}_m^T \mathbf{x} = 1$, such that we have the following:

$$\text{Primal feasibility:} \begin{cases} x_1^* + \cdots + x_m^* = 1 \\ x_i^* \geq 0 \quad i \in \{1, \ldots, m\}. \end{cases}$$

$$\text{Dual feasibility:} \begin{cases} \nabla_{\mathbf{x}} z(\mathbf{x}^*) - \sum_{i=1}^{m} \lambda_i(-\mathbf{e}_i) - \alpha \mathbf{1}_m = \mathbf{0} \\ \lambda_i \geq 0 \quad i \in \{1, \ldots, m\} \\ \alpha \quad \text{unrestricted.} \end{cases}$$

$$\text{Complementary slackness:} \{ \lambda_i x_i^* = 0 \quad i \in \{1, \ldots, m\}.$$

We observe first that $\nabla_{\mathbf{x}} z(\mathbf{x}) = \mathbf{A}\mathbf{y}^*$. Therefore, we can write the first equation in the dual feasibility condition as

$$\mathbf{A}\mathbf{y}^* - \alpha \mathbf{1}_m = -\sum_{i=1}^{m} \lambda_i \mathbf{e}_i. \tag{8.2}$$

Since $\lambda_i \geq 0$ and $\mathbf{e}_i$ is simply the ith standard basis vector, we know that $\lambda_i \mathbf{e}_i \geq \mathbf{0}$, and thus

$$\mathbf{A}\mathbf{y}^* - \alpha \mathbf{1}_m \leq \mathbf{0}. \tag{8.3}$$

Now, again consider the first equation in the dual feasibility condition written as

$$\mathbf{A}\mathbf{y}^* + \sum_{i=1}^{m} \lambda_i \mathbf{e}_i - \alpha \mathbf{1}_m = \mathbf{0}.$$

If we multiply this by $\mathbf{x}^{*T}$ on the left, we obtain

$$\mathbf{x}^{*T}\mathbf{A}\mathbf{y}^* + \sum_{i=1}^{m} \lambda_i \mathbf{x}^{*T} \mathbf{e}_i - \alpha \mathbf{x}^{*T} \mathbf{1}_m = \mathbf{x}^{*T}\mathbf{0}. \tag{8.4}$$

However, $\lambda_i \mathbf{x}^{*T} \mathbf{e}_i = \lambda_i x_i^* = 0$ by complementary slackness and $\alpha \mathbf{x}^{*T} \mathbf{1}_m = \alpha \mathbf{1}_m^T \mathbf{x}^* = \alpha$ by the primal feasibility conditions. Clearly, $\mathbf{x}^{*T}\mathbf{0} = 0$. Thus, we conclude from Eq. (8.4) that

$$\mathbf{x}^{*T}\mathbf{A}\mathbf{y}^* - \alpha = 0. \tag{8.5}$$

If we consider the problem for Player 2, then

$$z(y_1, \ldots, y_n) = z(\mathbf{y}) = \left(\mathbf{x}^{*T}\mathbf{B}\right)\mathbf{y} \tag{8.6}$$

so that the jth component of $\nabla_{\mathbf{y}} z(\mathbf{y})$ is $\mathbf{x}^{*T}\mathbf{B}_{.j}$. If we consider the KKT conditions for Player 2, we know that $\mathbf{y}^*$ is an optimal solution if and only if there exists Lagrange multipliers, $\mu_1, \ldots, \mu_n$, corresponding to the constraints $\mathbf{y} \geq \mathbf{0}$, and β, corresponding to the constraint $y_1 + \cdots + y_n = 1$, so that we have the following:

$$\text{Primal feasibility}: \begin{cases} y_1^* + \cdots + y_n^* = 1 \\ y_j^* \geq 0 \quad j \in \{1, \ldots, n\}. \end{cases}$$

$$\text{Dual feasibility}: \begin{cases} \nabla_{\mathbf{y}} z(\mathbf{y}^*) - \sum_{j=1}^{n} \mu_j(-\mathbf{e}_i) - \beta \mathbf{1}_n = \mathbf{0} \\ \mu_j \geq 0 \quad j \in \{1, \ldots, n\} \\ \beta \quad \text{unrestricted.} \end{cases}$$

Complementary slackness : $\{ \mu_j y_j^* = 0 \quad j \in \{1, \ldots, n\} \}$.

By the same arguments we used in the case for Player 1, we can show that

$$\mathbf{x}^{*T}\mathbf{B} - \beta \mathbf{1}_n^T \leq \mathbf{0}, \tag{8.7}$$

and

$$\mathbf{x}^{*T}\mathbf{B}\mathbf{y}^* - \beta = 0. \tag{8.8}$$

Thus, we have shown (from the necessity and sufficiency of KKT conditions for the two problems) that

$$\mathbf{x}^{*T}\mathbf{A}\mathbf{y}^* - \alpha = 0,$$
$$\mathbf{x}^{*T}\mathbf{B}\mathbf{y}^* - \beta = 0,$$
$$\mathbf{A}\mathbf{y}^* - \alpha \mathbf{1}_m \leq \mathbf{0},$$
$$\mathbf{x}^{*T}\mathbf{B} - \beta \mathbf{1}_n^T \leq \mathbf{0},$$
$$\mathbf{1}_m^T \mathbf{x}^* - 1 = 0,$$
$$\mathbf{1}_n^T \mathbf{y}^* - 1 = 0,$$
$$\mathbf{x}^* \geq \mathbf{0},$$
$$\mathbf{y}^* \geq \mathbf{0}$$

are necessary and sufficient conditions for $(\mathbf{x}^*, \mathbf{y}^*)$ to be a Nash equilibrium of the game $\mathcal{G}$. □

Theorem 8.12. *Let* $\mathcal{G} = (\mathbf{P}, \Sigma, \mathbf{A}, \mathbf{B})$ *be a general-sum, two-player bimatrix game, with matrices* $\mathbf{A}, \mathbf{B} \in \mathbb{R}^{m \times n}$. *A pair* $(\mathbf{x}^*, \mathbf{y}^*) \in \Delta$ *is* *a Nash equilibrium if and only if the tuple* $(\mathbf{x}^*, \mathbf{y}^*, \alpha^*, \beta^*)$ *is a global maximizer for the quadratic programming problem*

$$
\begin{aligned}
\max \ \ & \mathbf{x}^T(\mathbf{A} + \mathbf{B})\mathbf{y} - \alpha - \beta, \\
s.t. \ \ & \mathbf{A}\mathbf{y} - \alpha\mathbf{1}_m \leq \mathbf{0}, \\
& \mathbf{x}^T\mathbf{B} - \beta\mathbf{1}_n^T \leq \mathbf{0}, \\
& \mathbf{1}_m^T\mathbf{x} - 1 = 0, \\
& \mathbf{1}_n^T\mathbf{y} - 1 = 0, \\
& \mathbf{x} \geq \mathbf{0}, \\
& \mathbf{y} \geq \mathbf{0}.
\end{aligned}
\tag{8.9}
$$

Proof. Before proceeding to the proof, note that if we assume

$$
\mathbf{A}\mathbf{y} - \alpha\mathbf{1}_m \leq \mathbf{0},
$$

then multiplying both sides of the equation by $\mathbf{x}^T$ yields

$$
\mathbf{x}^T\mathbf{A}\mathbf{y} - \alpha\mathbf{x}^T\mathbf{1}_m \leq 0.
$$

Now, if $\mathbf{x} \in \Delta_m$, then $\mathbf{x}^T\mathbf{1}_m = \mathbf{1}_m^T\mathbf{x} = 1$, which implies that

$$
\mathbf{x}^T\mathbf{A}\mathbf{y} - \alpha \leq 0.
$$

Multiplying $\mathbf{x}^T\mathbf{B} - \beta\mathbf{1}_n^T \leq \mathbf{0}$ on the right by $\mathbf{y}$ yields a similar inequality:

$$
\mathbf{x}^T\mathbf{B}\mathbf{y} - \beta \leq 0.
$$

Adding these results together yields

$$
z(\mathbf{x}, \mathbf{y}, \alpha, \beta) = \mathbf{x}^T(\mathbf{A} + \mathbf{B})\mathbf{y} - \alpha - \beta \leq 0.
$$

Thus, any set of variables $(\mathbf{x}^*, \mathbf{y}^*, \alpha^*, \beta^*)$ so that $z(\mathbf{x}^*, \mathbf{y}^*, \alpha^*, \beta^*) = 0$ is a global maximum, assuming the constraints given in the theorem. Note also that the constraint

$$
\mathbf{A}\mathbf{y} - \alpha\mathbf{1}_m \leq \mathbf{0}
$$

is shorthand for m distinct constraints of the form

$$
\mathbf{A}_{i.}\mathbf{y} - \alpha \leq 0.
$$

Similarly, the constraint

$$\mathbf{x}^T\mathbf{B} - \beta\mathbf{1}_n^T \leq \mathbf{0}$$

is shorthand for n constraints of the form

$$\mathbf{x}^T\mathbf{B}_{\cdot j} - \beta \leq 0.$$

Finally, $\mathbf{x} \geq \mathbf{0}$ and $\mathbf{y} \geq \mathbf{0}$ are shorthand for $m + n$ additional constraints of the form $x_i \geq 0$, with $i \in \{1, \ldots, m\}$, and $y_j \geq 0$, with $j \in \{1, \ldots, n\}$. We use this information when we form KKT conditions. We now proceed to the proof.

($\Leftarrow$) We show that the KKT conditions for Eq. (8.9) are identical to the conditions given in Lemma 8.11. Thus, if $(\mathbf{x}^*, \mathbf{y}^*, \alpha^*, \beta^*)$ is a global optimum for Eq. (8.9), then it must satisfy the KKT conditions by Theorem 6.36; consequently, by Lemma 8.11, $(\mathbf{x}^*, \mathbf{y}^*)$ must be a Nash equilibrium for $\mathcal{G}$.

At an optimal point $(\mathbf{x}^*, \mathbf{y}^*, \alpha^*, \beta^*)$, there are the following multipliers:

(1) $\lambda_1, \ldots, \lambda_m$, corresponding to the constraints $\mathbf{A}\mathbf{y} - \alpha\mathbf{1}_m \leq \mathbf{0}$;
(2) $\mu_1, \ldots, \mu_n$, corresponding to the constraints $\mathbf{x}^T\mathbf{B} - \beta\mathbf{1}_n^T \leq \mathbf{0}$;
(3) ν_1, corresponding to the constraint $\mathbf{1}_m^T\mathbf{x} - 1$;
(4) ν_2, corresponding to the constraint $\mathbf{1}_n^T\mathbf{y} - 1 = 0$;
(5) $\phi_1, \ldots, \phi_m$, corresponding to the constraints $\mathbf{x} \geq \mathbf{0}$; and
(6) $\theta_1, \ldots, \theta_n$, corresponding to the constraints $\mathbf{y} \geq \mathbf{0}$.

Write $\mathbf{x} \geq \mathbf{0}$ as $-\mathbf{x} \leq \mathbf{0}$ and $\mathbf{y} \geq \mathbf{0}$ as $-\mathbf{y} \leq \mathbf{0}$. When computing the gradients of the various constraint and objective functions, note that we can compute the gradients of the various constraint functions and the objective function. Each gradient has $m + n + 2$ components: one for each variable in $\mathbf{x}$ and $\mathbf{y}$ and two additional components for α and β. Throughout the remainder of this proof, the vector $\mathbf{0}$ will vary in size to ensure that all vectors have the correct size. The gradients are:

(1)

$$\nabla z(\mathbf{x}, \mathbf{y}, \alpha, \beta) = \begin{bmatrix} (\mathbf{A} + \mathbf{B})\mathbf{y} \\ (\mathbf{A} + \mathbf{B})^T\mathbf{x} \\ -1 \\ -1 \end{bmatrix};$$

(2)

$$\nabla\left(\mathbf{A}_{i\cdot}\mathbf{y} - \alpha\right) = \begin{bmatrix} \mathbf{0} \\ \mathbf{A}_{i\cdot}^T \\ -1 \\ 0 \end{bmatrix} \quad i \in \{1, \ldots, m\};$$

(3)

$$\nabla\left(\mathbf{x}^T\mathbf{B}_{\cdot j} - \beta\right) = \begin{bmatrix} \mathbf{B}_{\cdot j} \\ \mathbf{0} \\ 0 \\ -1 \end{bmatrix} \quad j \in \{1, \ldots, n\};$$

(4)

$$\nabla(\mathbf{1}_m^T\mathbf{x} - 1) = \begin{bmatrix} \mathbf{1}_m \\ \mathbf{0} \\ 0 \\ 0 \end{bmatrix};$$

(5)

$$\nabla(\mathbf{1}_n^T\mathbf{y} - 1) = \begin{bmatrix} \mathbf{0} \\ \mathbf{1}_n \\ 0 \\ 0 \end{bmatrix};$$

(6)

$$\nabla(-x_i) = \begin{bmatrix} -\mathbf{e}_i \\ \mathbf{0} \\ 0 \\ 0 \end{bmatrix};$$

(7)

$$\nabla(-y_j) = \begin{bmatrix} \mathbf{0} \\ -\mathbf{e}_j \\ 0 \\ 0 \end{bmatrix}.$$

In the final two gradient computations, $\mathbf{e}_i \in \mathbb{R}^{m \times 1}$ and $\mathbf{e}_j \in \mathbb{R}^{n \times 1}$ so that the standard basis vectors agree with the dimensionality of $\mathbf{x}$ and $\mathbf{y}$, respectively. The dual feasibility constraints of the KKT conditions for the quadratic program assert that:

(1) $\lambda_1, \ldots, \lambda_n \geq 0$,
(2) $\mu_1, \ldots, \mu_m \geq 0$,
(3) $\phi_1, \ldots, \phi_m \geq 0$,
(4) $\theta_1, \ldots, \theta_n \geq 0$,
(5) $\nu_1 \in \mathbb{R}$, and
(6) $\nu_2 \in \mathbb{R}$.

The KKT equality that appears in the dual feasibility conditions is

$$
\underbrace{\begin{bmatrix} (\mathbf{A}+\mathbf{B})\mathbf{y} \\ (\mathbf{A}+\mathbf{B})^T\mathbf{x} \\ -1 \\ -1 \end{bmatrix}}_{\nabla z} - \sum_{i=1}^{m} \lambda_i \underbrace{\begin{bmatrix} \mathbf{0} \\ \mathbf{A}_{i\cdot}^T \\ -1 \\ 0 \end{bmatrix}}_{\nabla(\mathbf{A}_{i\cdot}\mathbf{y}-\alpha)} - \sum_{j=1}^{n} \mu_j \underbrace{\begin{bmatrix} \mathbf{B}_{\cdot j} \\ \mathbf{0} \\ 0 \\ -1 \end{bmatrix}}_{\nabla(\mathbf{x}^T\mathbf{B}_{\cdot j}-\beta)}
$$

$$
- \nu_1 \underbrace{\begin{bmatrix} \mathbf{1}_m \\ \mathbf{0} \\ 0 \\ 0 \end{bmatrix}}_{\nabla(\mathbf{1}_m^T\mathbf{x}-1)} - \nu_2 \underbrace{\begin{bmatrix} \mathbf{0} \\ \mathbf{1}_n \\ 0 \\ 0 \end{bmatrix}}_{\nabla(\mathbf{1}_n^T\mathbf{y}-1)} - \sum_{i=1}^{m} \phi_i \underbrace{\begin{bmatrix} \mathbf{e}_i \\ \mathbf{0} \\ 0 \\ 0 \end{bmatrix}}_{\nabla(-x_i)} - \sum_{j=1}^{n} \theta_j \underbrace{\begin{bmatrix} \mathbf{0} \\ -\mathbf{e}_j \\ 0 \\ 0 \end{bmatrix}}_{\nabla(-y_j)} = \mathbf{0}.
$$

$$(8.10)$$

We can analyze this expression row by row, starting with the last row. Writing the last row on its own yields

$$
-1 + \sum_{j=1}^{n} \mu_j = 0.
$$

This implies

$$
\sum_{j=1}^{n} \mu_j = 1.
$$

Similarly, analyzing the third row on its own yields

$$\sum_{i=1}^{m} \lambda_i = 1. \tag{8.11}$$

Thus, dual feasibility shows that $(\lambda_1, \ldots, \lambda_m) \in \Delta_m$ and $(\mu_1, \ldots \mu_n) \in \Delta_n$.

Rows 1 and 2 are more complex and represent m and n equations, corresponding to the m elements in $\mathbf{x}$ and the n elements in $\mathbf{y}$. Consider the jth component of the second row, which corresponds to the variable y_j. Isolating this component and simplifying yields the equation

$$\mathbf{x}^T \left(\mathbf{A}_{\cdot j} + \mathbf{B}_{\cdot j} \right) - \sum_{i=1}^{m} \lambda_i \mathbf{A}_{ij} - \nu_2 + \theta_j = 0.$$

Note that we have used a transpose on the first term to write everything in terms of the jth column of $\mathbf{A} + \mathbf{B}$, rather than the jth row of $(\mathbf{A} + \mathbf{B})^T$. In the remaining terms, we are simply extracting the jth component. This equation implies

$$\mathbf{x}^T \left(\mathbf{A}_{\cdot j} + \mathbf{B}_{\cdot j} \right) - \sum_{i=1}^{m} \lambda_i \mathbf{A}_{ij} - \nu_2 \leq 0. \tag{8.12}$$

We can similarly analyze the ith component of the first row corresponding to the variable x_i. The result is the (simplified) equation

$$(\mathbf{A}_{i \cdot} + \mathbf{B}_{i \cdot}) \mathbf{y} - \sum_{j=1}^{n} \mu_j \mathbf{B}_{ij} - \nu_1 \leq 0. \tag{8.13}$$

Eqs. (8.12) and (8.13) can be generalized to a system of equations (constraints) using matrix arithmetic. We have

$$\mathbf{x}^T (\mathbf{A} + \mathbf{B}) - \boldsymbol{\lambda}^T \mathbf{A} - \nu_2 \mathbf{1}_n^T \leq \mathbf{0} \tag{8.14}$$

since $j \in \{1, \ldots, n\}$, and

$$(\mathbf{A} + \mathbf{B}) \mathbf{y} - \mathbf{B} \boldsymbol{\mu} - \nu_1 \mathbf{1}_m \leq \mathbf{0} \tag{8.15}$$

since $i \in \{1, \ldots, m\}$. Here, $\boldsymbol{\lambda} = \langle \lambda_1, \ldots, \lambda_m \rangle$ and $\boldsymbol{\mu} = \langle \mu_1, \ldots, \mu_n \rangle$.

There is now a trick required to complete the proof. Suppose we choose the Lagrange multipliers so that $\boldsymbol{\lambda} = \mathbf{x}$ and $\boldsymbol{\mu} = \mathbf{y}$. We are free to do so because we have already proven that $\lambda \in \Delta_m$ and $\mu \in \Delta_n$. Furthermore, suppose we choose $\nu_1 = \alpha$ and $\nu_2 = \beta$. Then, if $\mathbf{x}^*$, $\mathbf{y}^*$, α^*, β^* is an optimal solution, then Eqs. (8.14) and (8.15) become

$$\mathbf{x}^{*T}(\mathbf{A} + \mathbf{B}) - \mathbf{x}^{*T}\mathbf{A} - \beta^* \mathbf{1}_n^T \leq \mathbf{0},$$

$$(\mathbf{A} + \mathbf{B})\mathbf{y}^* - \mathbf{B}\mathbf{y}^* - \alpha^* \mathbf{1}_m \leq \mathbf{0}.$$

Simplifying yields the inequalities

$$\mathbf{x}^{*T}\mathbf{B} - \beta^* \mathbf{1}_n^T \leq \mathbf{0},$$

$$\mathbf{A}\mathbf{y}^* - \alpha^* \mathbf{1}_m \leq \mathbf{0}.$$

We also know that:

(1) $\mathbf{1}_m^T \mathbf{x}^* = 1$,
(2) $\mathbf{1}_n^T \mathbf{y}^* = 1$,
(3) $\mathbf{x} \geq \mathbf{0}$, and
(4) $\mathbf{y} \geq \mathbf{0}$.

Finally, if we apply complementary slackness for the quadratic programming problem, we see that

$$\lambda_i \left(\mathbf{A}_{i\cdot}\mathbf{y} - \alpha\right) = 0 \quad i \in \{1, \ldots, m\},$$

$$\left(\mathbf{x}^T\mathbf{B}_{\cdot j} - \beta\right) \mu_j = 0 \quad j \in \{1, \ldots, n\}.$$

Using our assumptions that $\lambda_i = x_i^*$ and $\mu_j = y_j^*$ and summing over i yields

$$\sum_{i=1}^m x_i^* \left(\mathbf{A}_{i\cdot}\mathbf{y}^* - \alpha^*\right) = 0.$$

However, this implies

$$\sum_{i=1}^m x_i^* \mathbf{A}_{i\cdot}\mathbf{y}^* - \sum_{i=1}^m \alpha^* x_i^* = 0.$$

Rewriting this as a matrix equation yields

$$\mathbf{x}^{*T}\mathbf{A}\mathbf{y}^* - \alpha^* = 0. \tag{8.16}$$

Performing the same computation but this time summing over j yields

$$\sum_{j=1}^n \left(\mathbf{x}^{*T}\mathbf{B}_{\cdot j} - \beta^*\right) \mu_j = 0.$$

Following a similar analysis as before at last yields the expression

$$\mathbf{x}^{*T}\mathbf{B}\mathbf{y}^* - \beta^* = 0. \qquad (8.17)$$

If we add Eqs. (8.16) and (8.17), we obtain

$$\mathbf{x}^{*T}(\mathbf{A} + \mathbf{B})\mathbf{y}^* - \alpha^* - \beta^* = 0.$$

Thus, we have shown that any tuple, $(\mathbf{x}^*, \mathbf{y}^*, \alpha^*, \beta^*)$, satisfying the KKT conditions for the quadratic programming problem must also be a global maximizer because Section 8.3 asserts that the objective function of the quadratic programming problem takes on its maximum value. Moreover, by Lemma 8.11, it follows that such a point must also be a Nash equilibrium for $\mathcal{G}$.

($\Rightarrow$) The converse of the theorem states that if $(\mathbf{x}^*, \mathbf{y}^*)$ is a Nash equilibrium for $\mathcal{G}$, then setting $\alpha^* = \mathbf{x}^{*T}\mathbf{A}\mathbf{y}^*$ and $\beta^* = \mathbf{x}^{*T}\mathbf{B}\mathbf{y}^*$ gives an optimal solution, $(\mathbf{x}^*, \mathbf{y}^*, \alpha^*, \beta^*)$, to the quadratic program. It follows from Lemma 8.11 that when $(\mathbf{x}^*, \mathbf{y}^*)$ is a Nash equilibrium, we know that

$$\mathbf{x}^{*T}\mathbf{A}\mathbf{y}^* - \alpha^* = 0,$$
$$\mathbf{x}^{*T}\mathbf{B}\mathbf{y}^* - \beta^* = 0,$$

and thus we know at once that

$$\mathbf{x}^{*T}(\mathbf{A} + \mathbf{B})\mathbf{y}^* - \alpha^* - \beta^* = 0$$

holds. Therefore, it follows that $(\mathbf{x}^*, \mathbf{y}^*, \alpha^*, \beta^*)$ must be a global maximizer for the quadratic program because the objective function achieves its upper bound. This completes the proof. $\qquad\square$

Example 8.13. Recall the explicit Chicken game, which had the payoff matrices

$$\mathbf{A} = \begin{bmatrix} 0 & -1 \\ 1 & -10 \end{bmatrix},$$

$$\mathbf{B} = \begin{bmatrix} 0 & 1 \\ -1 & -10 \end{bmatrix}.$$

From Exercise 4.4, we know that the pure strategy $(\mathbf{e}_1, \mathbf{e}_2)$ is a Nash equilibrium. Symmetry tells us that the pure strategy $(\mathbf{e}_2, \mathbf{e}_1)$

must also be a Nash equilibrium. We can find a third, non-obvious, mixed-strategy Nash equilibrium using the quadratic programming formulation.

The quadratic program for this instance of Chicken is

$$
\begin{cases}
\max \quad \begin{bmatrix} x_1 & x_2 \end{bmatrix} \begin{bmatrix} 0 & 0 \\ 0 & -20 \end{bmatrix} \begin{bmatrix} y_1 \\ y_2 \end{bmatrix} - \alpha - \beta, \\[2ex]
\text{s.t.} \quad \begin{bmatrix} 0 & -1 \\ 1 & -10 \end{bmatrix} \begin{bmatrix} y_1 \\ y_2 \end{bmatrix} - \begin{bmatrix} \alpha \\ \alpha \end{bmatrix} \leq \begin{bmatrix} 0 \\ 0 \end{bmatrix}, \\[2ex]
\quad \begin{bmatrix} x_1 & x_2 \end{bmatrix} \begin{bmatrix} 0 & 1 \\ -1 & -10 \end{bmatrix} - \begin{bmatrix} \beta & \beta \end{bmatrix} \leq \begin{bmatrix} 0 & 0 \end{bmatrix}, \\[2ex]
\quad \begin{bmatrix} 1 & 1 \end{bmatrix} \begin{bmatrix} x_1 \\ x_2 \end{bmatrix} = 1, \\[2ex]
\quad \begin{bmatrix} 1 & 1 \end{bmatrix} \begin{bmatrix} y_1 \\ y_2 \end{bmatrix} = 1, \\[2ex]
\quad \begin{bmatrix} x_1 \\ x_2 \end{bmatrix} \geq \begin{bmatrix} 0 \\ 0 \end{bmatrix}, \\[2ex]
\quad \begin{bmatrix} y_1 \\ y_2 \end{bmatrix} \geq \begin{bmatrix} 0 \\ 0 \end{bmatrix}.
\end{cases} \tag{8.18}
$$

We can write this without matrix notation as

$$
\begin{cases}
\max \quad -20x_2y_2 - \alpha - \beta, \\
\text{s.t.} \quad -y_2 - \alpha \leq 0, \\
\quad y_1 - 10y_2 - \alpha \leq 0, \\
\quad -x_2 - \beta \leq 0, \\
\quad x_1 - 10x_2 - \beta \leq 0, \\
\quad x_1 + x_2 = 1, \\
\quad y_1 + y_2 = 1, \\
\quad x_1, x_2, y_1, y_2 \geq 0.
\end{cases} \tag{8.19}
$$

In Mathematica, the problem can be solved using the following code.

```
NMaximize[
    {
        -20x2*y2-a-b,              (*Objective *)
        -y2 - a <= 0,              (*Constraint 1*)
        y1 - 10y2 - a <= 0,        (*Constraint 2*)
        -x2 - b <= 0,              (*Constraint 3*)
        x1 - 10x2 - b <= 0,        (*Constraint 4*)
        x1 + x2 == 1,              (*Constraint 5*)
        y1 + y2 == 1,              (*Constraint 6*)
        x1>=0, x2>=0, y1>=0, y2>=0
    },
    {x1,x2,y1,y2,a,b} (*Variables*),
    Method -> "DifferentialEvolution" (*Method to use*)
]
```

We obtain the solution as

```
{0., {x1 -> 0.9, x2 -> 0.1,
    y1 -> 0.9, y2 -> 0.1,
    a -> -0.1, b -> -0.1}}.
```

That is, we have

$$x_1^* = y_1^* = \frac{9}{10} \quad x_2^* = y_2^* = \frac{1}{10}.$$

This is the mixed-strategy Nash equilibrium for this specific payoff matrix. We know that this must be a Nash equilibrium because the objective function correctly evaluates to zero at this point. It is possible (though a little challenging) to show that Chicken games always have the two pure Nash equilibria given above and a mixed-strategy Nash equilibrium, which can be found through this quadratic programming approach, if so desired.

8.4 Chapter Notes

Mangaserian and Stone's work [95], as presented in this chapter, was not the first to develop an algorithmic approach to computing

Nash equilibria in general-sum bimatrix games. Lemke and Howson [96] provided what is now called the Lemke–Howson algorithm and is generally considered more useful than the quadratic programming approach. However, the Lemke–Howson algorithm is more complex to describe and requires a custom implementation, making it less suitable for a first presentation. Lemke and Howson's approach yields an additional fact: Non-degenerate, two-player bimatrix games (where non-degenerate is defined in Ref. [96]) always have an odd number of Nash equilibria. This explains why the games we have seen so far seem to have one equilibrium or three equilibria. While it is relatively easy to construct degenerate games, this fact is not evident from the quadratic programming formulation. Wilson [97] generalized this result to N-player games, showing that in non-degenerate cases, they too have an odd number of equilibria. Lemke and Howson's approach has been further generalized to a theory of linear (and nonlinear) complementarity problems [98], which forms an entire class of mathematical programming problems. This has also led to methods for studying more general kinds of games, called Stackelberg games [99], using mixed complementarity approaches [100].

The computational methods developed for finding Nash equilibria, beginning with the work of Lemke and Howson [96] and Mangaserian and Stone [95], also mark the beginning of algorithmic game theory [3, 101–103], a topic of study in its own right. This area mixes computer science, mathematics, and economics to identify algorithmic foundations for analyzing games and finding equilibria. One of the main results to emerge from algorithmic game theory is the study of the computational complexity of finding a Nash equilibrium. Unlike linear programming problems, which we noted could be solved in polynomial time, arbitrary quadratic programming problems can be much harder to solve. In fact, it is possible to prove that solving an arbitrary quadratic programming problem is NP-hard, which is a formal way of classifying the difficulty of solving this problem. This would seem to settle the question of how difficult finding Nash equilibria is; however, it does not since we are not deriving an arbitrary quadratic programming problem but one with a specific form. Consequently, this has led to an exceptionally deep collection of results on the computational complexity of finding or approximating Nash equilibria. See Refs. [104–108] for examples.

Many of these results use a specialized computational complexity class called PPAD, introduced by Papadimitriou.

$$- \spadesuit \clubsuit \heartsuit \diamondsuit -$$

8.5 Exercises

8.1 Construct the quadratic programming problem for an instance of prisoner's dilemma. Show that the strategy in which both players defect is a Nash equilibrium.

8.2 Consider a Chicken game with the payoff matrices

$$\mathbf{A} = \begin{bmatrix} T & L \\ W & X \end{bmatrix}, \quad \mathbf{B} = \begin{bmatrix} T & W \\ L & X \end{bmatrix},$$

where $W > T > L > X$. Define

$$p = \frac{W - T}{(W - T) + (L - X)} \quad \gamma = \frac{LW - TX}{(W - T) + (L - X)}.$$

Show that $\mathbf{x} = \langle 1 - p, p \rangle$ and $\mathbf{y} = \langle 1 - p, p \rangle$ is the mixed-strategy Nash equilibrium for Chicken using $\alpha = \beta = \gamma$ in the quadratic programming formulation. If you like, you can derive this equilibrium yourself: Take the two constraints that come from $\mathbf{Ay} - \alpha \mathbf{1}_m \leq \mathbf{0}$, and subtract them to remove α. Treat the resulting inequality as an equation. Assuming $y_1 = 1 - p$ and $y_2 = p$, solve for p. You get the same result if you apply this to the constraints coming from $\mathbf{x}^T \mathbf{B} - \beta \mathbf{1}_n^T \leq \mathbf{0}$. Now, take the constraints and add them instead, using the values you just found for y_1 and y_2 (or x_1 and x_2). Treat this as an equation and solve for α (or β). You will obtain the formula that defines γ. It is also worth noting that this is not the simplest way to derive this equilibrium.

8.3 Show that when $\mathbf{B} = -\mathbf{A}$ (i.e., we have a zero-sum game), the quadratic programming problem reduces to the two dual linear programming problems we already identified in the previous chapter for solving zero-sum games.

Part 3
Cooperation in Game Theory

Chapter 9

Nash's Bargaining Problem and Cooperative Games

Chapter Goals: Heretofore, we have considered games in which the players were unable to communicate before play began or in which players have no way of trusting each other with certainty (remember prisoner's dilemma). In this chapter, we remove this restriction and consider those games in which players may put in place a pre-play agreement in an attempt to identify a solution with which both players can live happily. We begin by defining the cooperative expected payoff and payoff regions in both cooperative and competitive games. We then consider cooperative play as a multi-criteria optimization problem, defining the necessary machinery to formalize this, including Pareto optimality. We then introduce Nash's bargaining axioms and prove the Nash bargaining theorem using some tools from real analysis.

9.1 Payoff Regions in Two-Player Games

Remark 9.1. We will be dealing with bimatrix games throughout this chapter. We assume the two matrices have component forms:

$$\mathbf{A} = \begin{bmatrix} a_{11} & a_{12} & \cdots & a_{1n} \\ a_{21} & a_{22} & \cdots & a_{2n} \\ \vdots & \vdots & \ddots & \vdots \\ a_{m1} & a_{m2} & \cdots & a_{mn} \end{bmatrix}, \quad \mathbf{B} = \begin{bmatrix} b_{11} & b_{12} & \cdots & b_{1n} \\ b_{21} & b_{22} & \cdots & b_{2n} \\ \vdots & \vdots & \ddots & \vdots \\ b_{m1} & b_{m2} & \cdots & b_{mn} \end{bmatrix}.$$

Definition 9.2 (Cooperative Mixed Strategy). Let $\mathcal{G} = (\mathbf{P}, \Sigma, \mathbf{A}, \mathbf{B})$ be a two-player bimatrix game, with $\mathbf{A}, \mathbf{B} \in \mathbb{R}^{m \times n}$. A cooperative strategy is a collection of probabilities, x_{ij} ($i = 1, \ldots, m$, $j = 1, \ldots, n$), so that

$$\sum_{i=1}^{m} \sum_{j=1}^{n} x_{ij} = 1,$$

$$x_{ij} \geq 0 \quad i \in \{1, \ldots, m\}, \ j \in \{1, \ldots, n\}.$$

We associate a vector $\mathbf{x} \in \Delta_{mn}$ with this cooperative strategy.

Remark 9.3. For any cooperative strategy x_{ij} with $i \in \{1, \ldots, m\}$ and $j \in \{1, \ldots, n\}$, x_{ij} gives the probability that Player 1 plays row i while Player 2 plays column j. Note that $\mathbf{x}$ could be considered a matrix, but for the sake of notational consistency, it is easier to think of it as a vector with a strange indexing scheme.

Definition 9.4 (Cooperative Expected Payoff). Let $\mathcal{G} = (\mathbf{P}, \Sigma, \mathbf{A}, \mathbf{B})$ be a two-player bimatrix game, with $\mathbf{A}, \mathbf{B} \in \mathbb{R}^{n \times m}$, and let $\mathbf{x} \in \Delta_{mn}$ be a cooperative strategy for Players 1 and 2. Then,

$$u_1(\mathbf{x}) = \sum_{i=1}^{m} \sum_{j=1}^{n} a_{ij} x_{ij} \tag{9.1}$$

is the expected payoff for Player 1, while

$$u_2(\mathbf{x}) = \sum_{i=1}^{m} \sum_{j=1}^{n} b_{ij} x_{ij} \tag{9.2}$$

is the expected payoff for Player 2.

Definition 9.5 (Competitive Game Payoff Region). Let $\mathcal{G} = (\mathbf{P}, \Sigma, \mathbf{A}, \mathbf{B})$ be a two-player bimatrix game, with $\mathbf{A}, \mathbf{B} \in \mathbb{R}^{n \times m}$. The *payoff region of the competitive game* is

$$Q(\mathbf{A}, \mathbf{B}) = \{(u_1(\mathbf{x}, \mathbf{y}), u_2(\mathbf{x}, \mathbf{y})) : \mathbf{x} \in \Delta_m, \ \mathbf{y} \in \Delta_n\}, \tag{9.3}$$

where

$$u_1(\mathbf{x}, \mathbf{y}) = \mathbf{x}^T \mathbf{A} \mathbf{y},$$
$$u_2(\mathbf{x}, \mathbf{y}) = \mathbf{x}^T \mathbf{B} \mathbf{y}$$

are the standard player payoff functions.

Definition 9.6 (Cooperative Game Payoff Region). Let $\mathcal{G} = (\mathbf{P}, \Sigma, \mathbf{A}, \mathbf{B})$ be a two-player matrix game, with $\mathbf{A}, \mathbf{B} \in \mathbb{R}^{n \times m}$. The *payoff region of the cooperative game* is

$$P(\mathbf{A}, \mathbf{B}) = \{(u_1(\mathbf{x}), u_2(\mathbf{x})) : \mathbf{x} \in \Delta_{mn}\}, \tag{9.4}$$

where u_1 and u_2 are the cooperative payoff functions for Players 1 and 2, respectively.

Remark 9.7. The following lemma is left as an exercise.

Lemma 9.8. *Let* $\mathcal{G} = (\mathbf{P}, \Sigma, \mathbf{A}, \mathbf{B})$ *be a two-player matrix game, with* $\mathbf{A}, \mathbf{B} \in \mathbb{R}^{n \times m}$. *The competitive playoff region* $Q(\mathbf{A}, \mathbf{B})$ *is contained in the cooperative payoff region* $P(\mathbf{A}, \mathbf{B})$. □

Example 9.9. Consider a two-player bimatrix game with matrices

$$\mathbf{A} = \begin{bmatrix} 2 & -1 \\ -1 & 1 \end{bmatrix}, \quad \mathbf{B} = \begin{bmatrix} 1 & -1 \\ -1 & 2 \end{bmatrix}.$$

Games of this type are historically referred to as the Battle of the Sexes game, though in modern parlance they are called the Battle of the Buddies or, simply, coordination games. The story describes the decision-making process of a married couple or a pair of friends as they attempt to decide what to do on a given evening. In the classic Battle of the Sexes game, the players must decide whether to attend a boxing match or a ballet. One clearly prefers the boxing match (strategy 1 for each player), and the other prefers the ballet (strategy 2 for each player). Neither derives much benefit from going

(a) Competitive Region

(b) Cooperative Region

(c) Overlap

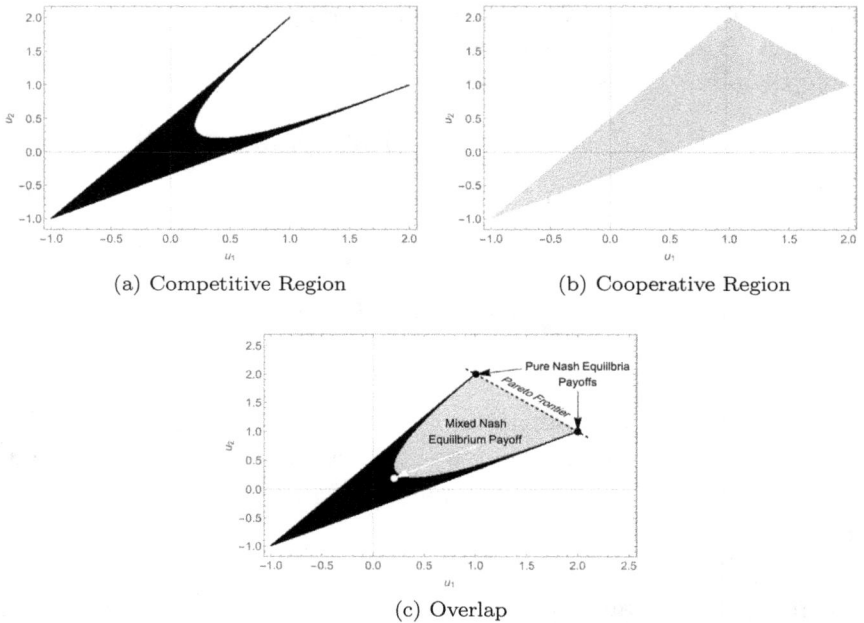

Fig. 9.1. The three plots show the competitive payoff region, the cooperative payoff region, and an overlay of the regions for the Battle of the Sexes game. Note that the cooperative payoff region completely contains the competitive payoff region.

to an event alone, which is indicated by the payoff of -1 in the off-diagonal elements. The competitive payoff region, the cooperative payoff region, and an overlay of the two regions for the Battle of the Sexes are shown in Fig. 9.1. Constructing these figures is done by brute force through in Mathematica.

Remark 9.10. Coordination games are generally described by payoff matrices in which the largest values occur on the diagonal, but in which players do not (necessarily) agree on the relative value of coordinated strategies. An example generalization of the example above would be a pair of payoff matrices,

$$\mathbf{A} = \begin{bmatrix} R & L \\ L & S \end{bmatrix}, \quad \mathbf{B} = \begin{bmatrix} S & L \\ L & R \end{bmatrix},$$

where $R > S > L$. Thus, Player 1 prefers it when both players play strategy 1, while Player 2 prefers it when both players play strategy 2 and neither prefers using opposite strategies. Thus, the players prefer to *coordinate* strategies. In contrast, an *anti-coordination game* has large values on the off-diagonal but smaller values on the diagonal.

Remark 9.11. Coordination games offer the possibility for cooperation since both players agree that coordination is preferred. The remainder of this chapter illustrates one way to build such a coordinated cooperative strategy.

Theorem 9.12. *Let* $\mathcal{G} = (\mathbf{P}, \Sigma, \mathbf{A}, \mathbf{B})$ *be a two-player bimatrix game, with* $\mathbf{A}, \mathbf{B} \in \mathbb{R}^{n \times m}$. *The cooperative payoff region* $P(\mathbf{A}, \mathbf{B})$ *is a convex set.*

Proof. The set $P(\mathbf{A}, \mathbf{B})$ is defined as a set of (u_1, u_2) satisfying the constraints

$$
\begin{cases}
\displaystyle\sum_{i=1}^{m}\sum_{j=1}^{n} a_{ij}x_{ij} - u_1 = 0, \\[2ex]
\displaystyle\sum_{i=1}^{m}\sum_{j=1}^{n} b_{ij}x_{ij} - u_2 = 0, \\[2ex]
\displaystyle\sum_{i=1}^{m}\sum_{j=1}^{n} x_{ij} = 1, \\[2ex]
x_{ij} \geq 0 \quad i \in \{1, \dots, m\},\ j \in \{1, \dots, n\}.
\end{cases} \tag{9.5}
$$

This set is defined by linear equalities (which are both convex and concave). Since linear functions are convex, the set of tuples $(u_1, u_2, \mathbf{x})$ that satisfy these constraints is a convex set by Theorems 6.18 and 6.29. Denote this set by X. It follows immediately that if $(u_1, u_2) \in P(\mathbf{A}, \mathbf{B})$, then there is a $\mathbf{x} \in \Delta_{mn}$ so that $(u_1, u_2, \mathbf{x}) \in X$.

Choose two points, $(u_1^1, u_2^1), (u_1^2, u_2^2) \in P(\mathbf{A}, \mathbf{B})$, and suppose that $(u_1^1, u_2^1, \mathbf{x}^1)$ and $(u_1^2, u_2^2, \mathbf{x}^2)$ are the corresponding elements of X. Since X is convex, for all $\lambda \in [0, 1]$, we have

$$
\lambda(u_1^1, u_2^1, \mathbf{x}^1) + (1 - \lambda)(u_1^2, u_2^2, \mathbf{x}^2) \in X. \tag{9.6}
$$

Let

$$
(u_1, u_2, \mathbf{x}) = \lambda(u_1^1, u_2^1, \mathbf{x}^1) + (1 - \lambda)(u_1^2, u_2^2, \mathbf{x}^2).
$$

But then, $(u_1, u_2) \in P(\mathbf{A}, \mathbf{B})$ because $(u_1, u_2, \mathbf{x}) \in X$. Therefore,

$$\lambda(u_1^1, u_2^1) + (1 - \lambda)(u_1^2, u_2^2) \in P(\mathbf{A}, \mathbf{B}) \qquad (9.7)$$

for all $\lambda \in [0, 1]$. It follows that $P(\mathbf{A}, \mathbf{B})$ is convex. $\qquad\square$

Remark 9.13. The following theorem assumes familiarity with the definition of a closed and bounded set in $\mathbb{R}^n$. A set is bounded in $\mathbb{R}^n$ if it can be enclosed in a ball (hypersphere) of sufficient size. There are many consistent definitions for a closed set in $\mathbb{R}^n$, all of which are outside the scope of this book and require a little topology or real analysis. As a consequence, we will not provide a complete proof of the following theorem but will take the whole statement as true. Those readers interested in a formal definition of closure (and compactness, which is what we are dealing with) might consult Munkres' text on topology [61] (Chapter 3).

Theorem 9.14. *Let* $\mathcal{G} = (\mathbf{P}, \Sigma, \mathbf{A}, \mathbf{B})$ *be a two-player bimatrix game, with* $\mathbf{A}, \mathbf{B} \in \mathbb{R}^{n \times m}$. *Let*

$$X = \left\{ (\mathbf{x}, u_1, u_2) \in \Delta_{mn} \times \mathbb{R} \times \mathbb{R} : \sum_{i=1}^{m} \sum_{j=1}^{n} a_{ij} x_{ij} - u_1 = 0, \right.$$

$$\left. \sum_{i=1}^{m} \sum_{j=1}^{n} b_{ij} x_{ij} - u_2 = 0 \right\}.$$

Then, both X *and the cooperative payoff region* $P(\mathbf{A}, \mathbf{B})$ *are bounded and closed sets.*

Sketch of Boundedness Proof. We note that X is equivalently defined by the expressions in Eq. (9.5). This set must be bounded because $x_{ij} \in [0, 1]$ for $i \in \{1, \ldots, m\}$ and $j \in \{1, \ldots, n\}$. As a result of this, the value of u_1 is bounded above and below by the largest and smallest values in $\mathbf{A}$, while the value of u_2 is bounded above and below by the largest and smallest values in $\mathbf{B}$. The fact that $P(\mathbf{A}, \mathbf{B})$ is bounded is clear from this argument as well. $\qquad\square$

Remark 9.15. We make use of this theorem when we analyze an optimization problem whose feasible region is X later in this chapter.

9.2 Collaboration and Multi-Criteria Optimization

Remark 9.16. Recall the generic optimization problem:

$$
\begin{cases}
\max \ z(x_1, \ldots, x_n), \\
\quad \text{s.t.} \ g_1(x_1, \ldots, x_n) \leq 0, \\
\qquad\qquad \vdots \\
\qquad\ g_m(x_1, \ldots, x_n) \leq 0, \\
\qquad\ h_1(x_1, \ldots, x_n) = 0, \\
\qquad\qquad \vdots \\
\qquad\ h_l(x_1, \ldots, x_n) = 0.
\end{cases}
$$

Here, $z : \mathbb{R}^n \to \mathbb{R}$, $g_i : \mathbb{R}^n \to \mathbb{R}$ $(i = 1, \ldots, m)$, and $h_j : \mathbb{R}^n \to \mathbb{R}$. This problem has a single objective function: $z(x_1, \ldots, x_n)$.

Definition 9.17. A *multi-criteria optimization problem* is an optimization problem with several simultaneous objective functions, $z_1, \ldots, z_s : \mathbb{R}^n \to \mathbb{R}$, written as

$$
\begin{cases}
\max \ \langle z_1(x_1, \ldots, x_n), z_2(x_1, \ldots, x_n), \cdots, z_s(x_1, \ldots, x_n) \rangle, \\
\quad \text{s.t.} \ g_1(x_1, \ldots, x_n) \leq 0, \\
\qquad\qquad \vdots \\
\qquad\ g_m(x_1, \ldots, x_n) \leq 0, \\
\qquad\ h_1(x_1, \ldots, x_n) = 0, \\
\qquad\qquad \vdots \\
\qquad\ h_l(x_1, \ldots, x_n) = 0.
\end{cases}
$$

Remark 9.18. Note that the objective function has now been replaced with a vector of objective functions. Multi-criteria optimization problems can be challenging to solve because, among other reasons, making $z_1(x_1, \ldots, x_n)$ larger may make $z_2(x_1, \ldots, x_n)$ smaller, and vice versa.

Example 9.19 (The Green Toy Maker). For the sake of argument, consider the toy maker problem from Example 7.3:

$$\begin{cases} \max \ z(x_1, x_2) = 7x_1 + 6x_2, \\ s.t. \ 3x_1 + x_2 \le 120, \\ \quad x_1 + 2x_2 \le 160, \\ \quad x_1 \le 35, \\ \quad x_1 \ge 0, \\ \quad x_2 \ge 0. \end{cases}$$

Suppose a certain amount of pollution is created each time a toy is manufactured. Each plane generates 3 units of pollution during manufacturing, while each boat generates only 2 units of pollution. Since x_1 is the number of planes produced and x_2 is the number of boats produced, we can create a multi-criteria optimization problem in which we simultaneously attempt to maximize profit $7x_1 + 6x_2$ and minimize pollution $3x_1 + 2x_2$. Since every minimization problem can be transformed into a maximization problem by negating the objective, we have the problem

$$\begin{cases} \max \ \langle 7x_1 + 6x_2, -3x_1 - 2x_2 \rangle, \\ s.t. \ 3x_1 + x_2 \le 120, \\ \quad x_1 + 2x_2 \le 160, \\ \quad x_1 \le 35, \\ \quad x_1 \ge 0, \\ \quad x_2 \ge 0. \end{cases}$$

Remark 9.20. For $n > 1$, we can choose many ways to order elements in $\mathbb{R}^n$. For example, on the plane, there are many ways to decide that a point (x_1, y_1) is greater than, less than, or equivalent to another point (x_2, y_2). We can think of these as the various ways of assigning a preference relation $\succ$ to points on the plane (or more generally points in $\mathbb{R}^n$). Here are three common ways to assign an order to points on the plane:

(1) Points can be ordered based on their Euclidean distance to the origin, i.e.,

$$(x_1, y_1) \succ (x_2, y_2) \iff \sqrt{x_1^2 + y_1^2} > \sqrt{x_2^2 + y_2^2}.$$

(2) A lexicographic ordering can be used by comparing the first component and then the second component.
(3) A parameter, $\lambda \in \mathbb{R}$, can be specified, and the ordering from $\mathbb{R}$ can be used via the formula

$$(x_1, y_1) \succ (x_2, y_2) \iff x_1 + \lambda y_1 > x_2 + \lambda y_2.$$

For this reason, a multi-criteria optimization problem may have many equally good solutions, depending on the order chosen. The following definition gives us a way to think about the tradeoffs that may occur for various ordering choices.

Definition 9.21 (Pareto Optimality). Let $g_i : \mathbb{R}^n \to \mathbb{R}$ ($i = 1, \ldots, m$), $h_j : \mathbb{R}^n \to \mathbb{R}$, and $z_k : \mathbb{R}^n \to \mathbb{R}$ ($k = 1, \ldots, s$). Consider the multi-criteria optimization problem

$$\begin{cases} \max & \langle z_1(x_1, \ldots, x_n), z_2(x_1, \ldots, x_n), \cdots, z_s(x_1, \ldots, x_n) \rangle, \\ s.t. & g_1(x_1, \ldots, x_n) \leq 0, \\ & \quad \vdots \\ & g_m(x_1, \ldots, x_n) \leq 0, \\ & h_1(x_1, \ldots, x_n) = 0, \\ & \quad \vdots \\ & h_l(x_1, \ldots, x_n) = 0. \end{cases}$$

A payoff vector, $\mathbf{z}(\mathbf{x}^*)$, dominates another payoff vector, $\mathbf{z}(\mathbf{x})$ (for two feasible points $\mathbf{x}, \mathbf{x}^*$), if:

(1) $\mathbf{z}_k(\mathbf{x}^*) \geq \mathbf{z}_k(\mathbf{x})$ for all $k \in \{1, \ldots, s\}$ and
(2) $\mathbf{z}_k(\mathbf{x}^*) > \mathbf{z}_k(\mathbf{x})$ for at least one $k \in \{1, \ldots, s\}$.

A solution, $\mathbf{x}^*$, is said to be *Pareto optimal* if $\mathbf{z}(\mathbf{x}^*)$ is not dominated by any other vector, $\mathbf{z}(\mathbf{x})$, where $\mathbf{x}$ is any other feasible solution.

Remark 9.22. A solution, $\mathbf{x}^*$, is Pareto optimal if changing the strategy can only benefit one objective function at the expense of another objective function. Put in terms of Example 9.19, a production pattern (x_1^*, x_2^*) is Pareto optimal if there is no way to change either x_1 or x_2 to both increase profit *and* decrease pollution.

Definition 9.23 (Multi-criteria Optimization Problem for Cooperative Games). Let $\mathcal{G} = (\mathbf{P}, \Sigma, \mathbf{A}, \mathbf{B})$ be a two-player matrix game, with $\mathbf{A}, \mathbf{B} \in \mathbb{R}^{n \times m}$. Then, the cooperative game multi-criteria optimization problem is

$$\begin{cases} \max \quad \langle u_1(\mathbf{x}) - u_1^0, u_2(\mathbf{x}) - u_2^0 \rangle, \\ \\ s.t. \quad \displaystyle\sum_{i=1}^{m}\sum_{j=1}^{n} x_{ij} = 1, \\ \\ \quad x_{ij} \geq 0 \quad i \in \{1, \ldots, m\} \ j \in \{1, \ldots, n\}. \end{cases} \quad (9.8)$$

Here, $\mathbf{x}$ is a cooperative mixed strategy;

$$u_1(\mathbf{x}) = \sum_{i=1}^{m}\sum_{j=1}^{n} a_{ij} x_{ij},$$

$$u_2(\mathbf{x}) = \sum_{i=1}^{m}\sum_{j=1}^{n} b_{ij} x_{ij}$$

are the cooperative expected payoff functions; and u_1^0 and u_2^0 are *status quo* payoff values – usually assumed to be Nash equilibrium payoff values for the two players.

Example 9.24. The three Nash equilibria of the Battle of the Sexes game, along with the Pareto optimal payoff points, are illustrated in Fig. 9.2. Note that in a maximization problem, the Pareto payoff points are always *up and to the right* and are at the boundary of the region containing all the possible payoffs. In a sense, this makes these points very similar to the solutions of linear programming problems. The set of points that are all Pareto optimal is sometimes called the *Pareto frontier.*

Remark 9.25. There is a substantial amount of research on solving multi-criteria optimization problems, which arise frequently in the real world. The interested reader might consider Ref. [109].

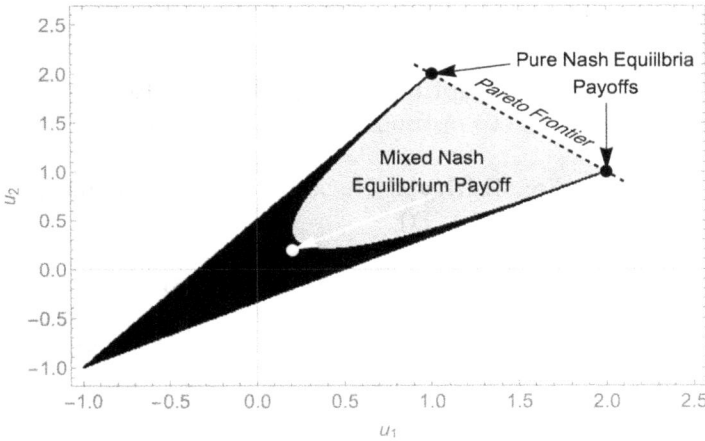

Fig. 9.2. The Pareto payoff points in $\mathbf{P}(\mathbf{A}, \mathbf{B})$, along with the payoff points of the three Nash equilibria of the Battle of the Sexes.

9.3 Nash's Bargaining Axioms

Remark 9.26. For a two-player matrix game, $\mathcal{G} = (\mathbf{P}, \Sigma, \mathbf{A}, \mathbf{B})$, with $\mathbf{A}, \mathbf{B} \in \mathbb{R}^{n \times m}$, Nash studied the problem of finding a cooperative mixed strategy, $\mathbf{x} \in \Delta_{mn}$, that would maximally benefit both players. The resulting strategy $\mathbf{x}^*$ is referred to as an *arbitration procedure* and is agreed to by the two players before play begins. In essence, we imagine this as a pre-play contract that can be enforced by charging a large penalty if either player breaks the contract to return the Nash equilibrium.

Remark 9.27. In solving this problem, Nash quantified six axioms (or assumptions) that he wished to ensure. We describe them as follows.

Assumption 1 (Rationality). If $\mathbf{x}^*$ is an arbitration procedure, we must have $u_1(\mathbf{x}^*) \geq u_1^0$ and $u_2(\mathbf{x}^*) \geq u_2^0$, where $(u_1^0, u_2^0) \in Q(A, B)$ is a non-cooperative status quo, usually a Nash equilibrium payoff.

Remark 9.28. Assumption 1 asserts that no player should be incentivized to return to the non-cooperative status quo.

Assumption 2 (Pareto Optimality). Any arbitration procedure, $\mathbf{x}^*$, is a Pareto-optimal solution to the two-player cooperative game multi-criteria optimization problem in Eq. (9.8). That is, $(u_1(\mathbf{x}^*), u_2(\mathbf{x}^*)$ is Pareto optimal.

Assumption 3 (Feasibility). The arbitration procedure $\mathbf{x}^* \in \Delta_{mn}$ and $(u_1(\mathbf{x}^*), u_2(\mathbf{x}^*) \in P(\mathbf{A}, \mathbf{B})$.

Assumption 4 (Independence of Irrelevant Alternatives). If $\mathbf{x}^*$ is an arbitration procedure and $P' \subseteq P(\mathbf{A}, \mathbf{B})$, with $(u_1^0, u_2^0), (u_1(\mathbf{x}^*), u_2(\mathbf{x}^*)) \in P'$, then $\mathbf{x}^*$ is still an arbitration procedure when we restrict our attention to P' (and the corresponding subset of Δ_{mn}).

Remark 9.29. Assumption 4 may seem odd. It was constructed to deal with restrictions on the payoff space, which in turn result in a restriction on the space of feasible solutions to the two-player cooperative game multi-criteria optimization problem. It states that if we add constraints to Eq. (9.8) but the resulting multi-criteria problem still has u_1^0 and u_2^0 as valid status quo values and the current arbitration procedure is still available because $(u_1(\mathbf{x}^*), u_2(\mathbf{x}^*), \mathbf{x}^*)$ remains in the reduced feasible region, then the arbitration procedure will not change, even though we have restricted the feasible region.

Assumption 5 (Invariance Under Linear Transformation). If $u_1(\mathbf{x})$ and $u_2(\mathbf{x})$ are replaced by $u_i'(\mathbf{x}) = \alpha_i u_i(\mathbf{x}) + \beta_i$ and $\alpha_i > 0$ and $u_i^{0'} = \alpha_i u_i^0 + \beta_i$ for $(i \in \{1, 2\})$, respectively, and $\mathbf{x}^*$ is an arbitration procedure for the original problem, then it is also an arbitration procedure for the transformed problem defined in terms of u_i' and $u_i^{0'}$.

Remark 9.30. Assumption 5 states that arbitration procedures are not affected by linear transformations of the underlying (linear) utility function. (See Theorem 2.27.)

Definition 9.31 (Symmetry of $P(\mathbf{A}, \mathbf{B})$). Let $\mathcal{G} = (\mathbf{P}, \Sigma, \mathbf{A}, \mathbf{B})$ be a two-player matrix game, with $\mathbf{A}, \mathbf{B} \in \mathbb{R}^{n \times m}$. The set $P(\mathbf{A}, \mathbf{B})$ is symmetric if, whenever $(u_1, u_2) \in P(\mathbf{A}, \mathbf{B})$, then $(u_2, u_1) \in P(\mathbf{A}, \mathbf{B})$.

Assumption 6 (Symmetry). If $P(\mathbf{A}, \mathbf{B})$ is symmetric and $u_1^0 = u_2^0$, then the arbitration procedure $\mathbf{x}^*$ has the property that $u_1(\mathbf{x}^*) = u_2(\mathbf{x}^*)$.

Remark 9.32. Assumption 6 states that if $P(\mathbf{A}, \mathbf{B})$ is symmetric, i.e., $P(\mathbf{A}, \mathbf{B})$ is symmetric in $\mathbb{R}^2$ about the line $y = x$, then both players will obtain the same payoff in an arbitration procedure, $\mathbf{x}^*$. An inspection of Fig. 9.1 reveals that this is a symmetric payoff region.

Remark 9.33. Our goal is to show that there is an arbitration procedure, $\mathbf{x}^* \in \Delta_{nm}$, that satisfies these assumptions and that the resulting pair $(u_1(\mathbf{x}^*), u_2(\mathbf{x}^*)) \in P(\mathbf{A}, \mathbf{B})$ is unique. This is *Nash's bargaining theorem*.

9.4 Nash's Bargaining Theorem

Remark 9.34. We begin our proof of Nash's bargaining theorem with two lemmas. We will not prove the first, as it requires a bit of analysis. The interested reader can refer to Rudin's little book on analysis [67].

Lemma 9.35 (Weierstrass' Theorem). *Let X be a non-empty closed and bounded set in $\mathbb{R}^n$, and let z be a continuous mapping with $z : S \to \mathbb{R}$. Then, the optimization problem*

$$
\begin{cases}
\max \ z(\mathbf{x}), \\
s.t. \ \mathbf{x} \in X,
\end{cases}
\tag{9.9}
$$

has at least one solution: $\mathbf{x}^* \in X$. $\qquad\square$

Remark 9.36. The final lemma sets the stage for Nash's bargaining theorem. Note that, in this lemma, we choose an explicit method of changing the multi-criteria problem in Definition 9.23 to a single-criterion optimization problem. That is, we choose a way to order pairs of real numbers. We also note that the proof of this result is rather long. On a first reading, it is fine to skip the proof or simply read the existence portion. Also, this proof is adapted from the one in Ref. [2], with some simplifications.

Lemma 9.37. *Let $\mathcal{G} = (\mathbf{P}, \Sigma, \mathbf{A}, \mathbf{B})$ be a two-player matrix game, with $\mathbf{A}, \mathbf{B} \in \mathbb{R}^{n \times m}$. Let $(u_1^0, u_2^0) \in P(\mathbf{A}, \mathbf{B})$. The quadratic programming problem*

$$
\begin{cases}
\max \ (u_1 - u_1^0)(u_2 - u_2^0), \\[2mm]
s.t. \ \displaystyle\sum_{i=1}^{m}\sum_{j=1}^{n} a_{ij} x_{ij} - u_1 = 0, \\[4mm]
\displaystyle\sum_{i=1}^{m}\sum_{j=1}^{n} b_{ij} x_{ij} - u_2 = 0, \\[4mm]
\displaystyle\sum_{i=1}^{m}\sum_{j=1}^{n} x_{ij} = 1, \\[4mm]
x_{ij} \geq 0 \quad i \in \{1, \ldots, m\}, \ j \in \{1, \ldots, n\}, \\[2mm]
u_1 \geq u_1^0, \\[2mm]
u_2 \geq u_2^0
\end{cases}
$$

has at least one global optimal solution: $(u_1^*, u_2^*, \mathbf{x}^*)$. Furthermore, if $(u_1', u_2', \mathbf{x}')$ is an alternative optimal solution, then $u_1^* = u_1'$ and $u_2^* = u_2'$.

Proof.   As before, denote the feasible region by X. By the same argument as in the proof of Theorem 9.14, X is a closed, bounded set. Moreover, since $(u_1^0, u_2^0) \in P(\mathbf{A}, \mathbf{B})$, we know that there is some $\mathbf{x}^0$ satisfying the constraints given in Eq. (9.5) and that the tuple $(u_1^0, u_2^0, \mathbf{x}^0)$ is feasible for this problem. Thus, the feasible region is non-empty. Applying Lemma 9.35, we know that there is at least one (global optimal) solution to this problem.

To see the uniqueness of (u_1^*, u_2^*), suppose that $V = (u_1^* - u_1^0)(u_2^* - u_2^0)$, and we have a second solution, $(u_1', u_2', \mathbf{x}')$, so that (without loss of generality) $u_1' > u_1^*$ and $u_2' < u_2^*$ but $V = (u_1' - u_1^0)(u_2' - u_2^0)$. By Theorem 9.12, we know X is convex; therefore,

$$
(\bar{u}_1, \bar{u}_2, \bar{\mathbf{x}}) = \frac{1}{2}(u_1^*, u_2^*, \mathbf{x}^*) + \frac{1}{2}(u_1', u_2', \mathbf{x}') \in X.
$$

Evaluate the objective function at this point to obtain

$$
\bar{V} = (\bar{u}_1 - u_1^0)(\bar{u}_2 - u_2^0) = \left(\frac{1}{2}u_1^* + \frac{1}{2}u_1' - u_1^0\right)\left(\frac{1}{2}u_2^* + \frac{1}{2}u_2' - u_2^0\right).
$$

Here, we are simply defining $\bar{V}$ as the value of the objective function at the new point. Note that for $i \in \{1, 2\}$, we can write

$$\frac{1}{2}u_i^* + \frac{1}{2}u_i' - u_i^0 = \frac{1}{2}(u_i^* - u_i^0) + \frac{1}{2}(u_i' - u_i^0)$$

$$= \frac{1}{2}\left((u_i^* - u_i^0) + (u_i' - u_i^0)\right).$$

Then, the objective function at the new point becomes

$$\bar{V} = \frac{1}{4}\left((u_1^* - u_1^0) + (u_1' - u_1^0)\right)\left((u_2^* - u_2^0) + (u_2' - u_2^0)\right).$$

Our objective is to show that this is an improved objective value. Let

$$Q_i^* = u_i^* - u_i^0 \quad Q_i' = u_i' - u_i^0.$$

Using these expressions, we can write

$$\bar{V} = \frac{1}{4}\left(Q_1^* + Q_1'\right)\left(Q_2^* + Q_2'\right).$$

Before proceeding, note that from our definition of V and our assumptions on u_1' and u_2', we have

$$V = Q_1^* Q_2^* = Q_1' Q_2'.$$

Using this information, expand the expression for $\bar{V}$ to obtain

$$\bar{V} = \frac{1}{4}\left(\underbrace{Q_1^* Q_2^*}_{V} + \underbrace{Q_1' Q_2'}_{V} + Q_1^* Q_2' + Q_1' Q_2^*\right)$$

$$= \frac{1}{4}\left(2V + Q_1^* Q_2' + Q_1' Q_2^*\right). \tag{9.10}$$

The next part of the proof requires an algebraic trick. Expanding Q_i^* and Q_i' in u_i^* and u_i' for $i \in \{1, 2\}$ and applying some arithmetic[1] will show that the following equality holds:

$$Q_1^* Q_2' + Q_1' Q_2^* - 2V = Q_1^* Q_2' + Q_1' Q_2^* - Q_1^* Q_2^* - Q_1' Q_2'$$

$$= (u_1^* - u_1')(u_2' - u_2^*).$$

---

[1]You can check this by hand or use a computer algebra system, such as Mathematica^{TM}.

We initially assumed that $u_1^* > u_1'$ and $u_2' > u_2^*$; therefore,

$$Q_1^* Q_2' + Q_1' Q_2^* - 2V > 0.$$

Let $R = Q_1^* Q_2' + Q_1' Q_2^* - 2V$. Assembling the pieces, we can rewrite Eq. (9.10) as

$$\bar{V} = \frac{1}{4} \left(4V + Q_1^* Q_2' + Q_1' Q_2^* - 2V \right) = \frac{1}{4} \left(4V + R \right) = V + \frac{R}{4} > V$$

because $R > 0$. However, this contradicts our assumption that $(u_1^*, u_2^*, \mathbf{x}^*)$ and $(u_1', u_2', \mathbf{x}')$ were optimal solutions to the problem because $\bar{V}$ is an improved objective value. Therefore, we must have $u_1^* = u_1'$ and $u_2^* = u_2'$. This completes the proof. $\qquad \square$

Theorem 9.38 (Nash' Bargaining Theorem). *Let* $\mathcal{G} = (\mathbf{P}, \Sigma, \mathbf{A}, \mathbf{B})$ *be a two-player matrix game, with* $\mathbf{A}, \mathbf{B} \in \mathbb{R}^{m \times n}$ *and with* $(u_1^0, u_2^0) \in P(\mathbf{A}, \mathbf{B})$. *Then, there is at least one arbitration procedure,* $\mathbf{x}^* \in \Delta_{mn}$, *satisfying the six bargaining assumptions and, moreover, the payoffs* $u_1(\mathbf{x}^*)$ *and* $u_2(\mathbf{x}^*)$ *are the unique payoff optimal points in* $P(\mathbf{A}, \mathbf{B})$.

Proof. Consider the quadratic programming problem from Lemma 9.37:

$$
\begin{cases}
\max \ (u_1 - u_1^0)(u_2 - u_2^0), \\[2mm]
s.t. \ \displaystyle\sum_{i=1}^{m} \sum_{j=1}^{n} a_{ij} x_{ij} - u_1 = 0, \\[4mm]
\displaystyle\sum_{i=1}^{m} \sum_{j=1}^{n} b_{ij} x_{ij} - u_2 = 0, \\[4mm]
\displaystyle\sum_{i=1}^{m} \sum_{j=1}^{n} x_{ij} = 1, \\[4mm]
x_{ij} \geq 0 \quad i = 1, \ldots, m, \ j = 1, \ldots, n, \\[2mm]
u_1 \geq u_1^0, \\[2mm]
u_2 \geq u_2^0.
\end{cases}
\qquad (9.11)
$$

It suffices to show that the solution to this quadratic program provides an arbitration procedure, $\mathbf{x}$, satisfying Nash's assumptions.

Uniqueness follows immediately from Lemma 9.37. As before, denote the feasible region of this problem by X.

Before proceeding, recall that $Q(\mathbf{A}, \mathbf{B})$, the payoff region for the competitive game $\mathcal{G}$, is contained in $P(\mathbf{A}, \mathbf{B})$. If u_1^0, u_2^0 is chosen as an equilibrium for the competitive game, we know that $(u_1^0, u_2^0) \in P(\mathbf{A}, \mathbf{B})$. Thus, there is a $\mathbf{x}^0$ so that $(u_1^0, u_2^0, \mathbf{x}^0) \in X$, and it follows that 0 is a lower bound for the maximal value of the objective function. We now prove that the Nash bargaining assumptions hold.

Assumption 1. By the construction of this problem, we know that $u_1(\mathbf{x}^*) \geq u_1^0$ and $u_2(\mathbf{x}^*) \geq u_2^0$.

Assumption 2. By Lemma 9.37, any solution, $(u_1^*, u_2^*, \mathbf{x}^*)$, has unique u_1^* and u_2^*. Thus, any other feasible solution, $(u_1, u_2, \mathbf{x})$, must have the property that either $u_1 < u_1^*$ or $u_2 < u_2^*$. Therefore, (u_1^*, u_2^*) must be Pareto optimal.

Assumption 3. Since the constraints of Eq. (9.11) properly contain the constraints in Eq. (9.5), the assumption of feasibility is ensured.

Assumption 4. Suppose that $P' \subseteq P(\mathbf{A}, \mathbf{B})$. Then, there is a subset $X' \subseteq X$ corresponding to P'. If both $(u_1^*, u_2^*) \in P'$ and $(u_1^0, u_2^0) \in P'$, it follows that there is a pair of cooperative mixed strategies, $\mathbf{x}^*, \mathbf{x}^0 \in \Delta_{mn}$, so that $(u_1^*, u_2^*, \mathbf{x}^*) \in X'$ and $(u_1^0, u_2^0, \mathbf{x}^0) \in X'$. Then, we can define the new optimization problem

$$
\begin{cases}
\max \ (u_1 - u_1^0)(u_2 - u_2^0), \\
\text{s.t.} \ (u_1, u_2, \mathbf{x}) \in X, \\
\quad\quad (u_1, u_2, \mathbf{x}) \in X'.
\end{cases} \tag{9.12}
$$

These constraints are consistent, and since

$$
(u_1^* - u_1^0)(u_2^* - u_2^0) \geq (u_1' - u_1^0)(u_2' - u_2^0) \tag{9.13}
$$

for all $(u_1', u_2', \mathbf{x}') \in X$, it follows that Eq. (9.13) must also hold for all $(u_1', u_2', \mathbf{x}') \in X' \subseteq X$. Thus, $(u_1^*, u_2^*, \mathbf{x}^*)$ is also an optimal solution for Eq. (9.12).

Assumption 5. Replace the objective function in Eq. (9.11) with

$$
\left(\alpha_1 u_1 + \beta_1 - (\alpha_1 u_1^0 - \beta_1) \right) \left(\alpha_2 u_2 + \beta_2 - (\alpha_2 u_2^0 - \beta_2) \right)
$$
$$
= \alpha_1 \alpha_2 (u_1 - u_1^0)(u_2 - u_2^0). \tag{9.14}
$$

The constraints of the problem will not be changed since we assume that $\alpha_1, \alpha_2 \geq 0$. To see this, note that a linear transformation of the payoff values implies the new constraints

$$\sum_{i=1}^{m} \sum_{j=1}^{n} (\alpha_1 a_{ij} + \beta_1) x_{ij} - (\alpha_1 u_1 + \beta_1) = 0 \iff$$

$$\alpha_1 \sum_{i=1}^{m} \sum_{j=1}^{n} a_{ij} x_{ij} + \beta_1 \underbrace{\sum_{i=1}^{m} \sum_{j=1}^{n} x_{ij}}_{\text{Simplifies to 1}} - (\alpha_1 u_1 + \beta_1) = 0 \iff$$

$$\alpha_1 \sum_{i=1}^{m} \sum_{j=1}^{n} a_{ij} x_{ij} + \beta_1 - \alpha_1 u_1 - \beta_1 = 0 \iff$$

$$\sum_{i=1}^{m} \sum_{j=1}^{n} a_{ij} x_{ij} - u_1 = 0. \quad (9.15)$$

A similar result holds for the constraints

$$\sum_{i=1}^{m} \sum_{j=1}^{n} b_{ij} x_{ij} - u_2 = 0.$$

Since the constraints are identical, it is clear that changing the objective function to the function in Eq. (9.14) will not affect the solution since we are simply scaling the value by a positive number.

Assumption 6. Suppose that $u^0 = u_1^0 = u_2^0$ and $P(\mathbf{A}, \mathbf{B})$ is symmetric. Assuming that $P(\mathbf{A}, \mathbf{B})$ is symmetric implies that $(u_2^*, u_1^*) \in P(\mathbf{A}, \mathbf{B})$ and that

$$(u_1^* - u_1^0)(u_2^* - u_2^0) = (u_1^* - u^0)(u_2^* - u^0) = (u_2^* - u^0)(u_1^* - u^0). \quad (9.16)$$

Thus, for some $\mathbf{x}' \in \Delta_{mn}$. we know that $(u_2^*, u_1^*, \mathbf{x}') \in X$ since $(u_2^*, u_1^*) \in P(\mathbf{A}, \mathbf{B})$. However, this feasible solution achieves the same objective value as the optimal solution, $(u_1^*, u_2^*, \mathbf{x}^*) \in X$, and thus by Lemma 9.37, we know that $u_1^* = u_2^*$.

Thus, we have shown that the six Nash bargaining assumptions are satisfied. This completes the proof. $\qquad\square$

Example 9.39. Consider the Battle of the Sexes game. Recall that

$$\mathbf{A} = \begin{bmatrix} 2 & -1 \\ -1 & 1 \end{bmatrix}, \quad \mathbf{B} = \begin{bmatrix} 1 & -1 \\ -1 & 2 \end{bmatrix}.$$

We can now find the arbitration process that produces the best cooperative strategy for the two players. This game has three Nash equilibria. The first two are in pure strategies:

(1) $(\mathbf{e}_2, \mathbf{e}_1)$: In this equilibrium, Player 2 always gets to go to the ballet and Player 1 never goes to the boxing match.
(2) $(\mathbf{e}_1, \mathbf{e}_2)$: In this equilibrium, Player 1 always gets to go to the boxing match and Player 1 never goes to the ballet.

The third equilibrium is a mixed-strategy equilibrium. We leave the details of finding this equilibrium as Exercise 9.2. However, the payoff to the two players at this equilibrium are $u_1^0 = u_2^0 = \frac{1}{5}$. Note that the two pure strategy equilibria are unfair, while the mixed-strategy equilibrium yields low total happiness.

We can apply Nash bargaining to try to improve the scenario. Assume that we begin with the mixed-strategy Nash equilibrium. Using the status quo, $u_1^0 = u_2^0 = 1/5$, the problem we must solve is

$$\max \left(u_1 - \frac{1}{5} \right) \left(u_2 - \frac{1}{5} \right),$$

$$s.t. \ 2x_{11} - x_{12} - x_{21} + x_{22} - u_1 = 0,$$

$$x_{11} - x_{12} - x_{21} + 2x_{22} - u_2 = 0,$$

$$x_{11} + x_{12} + x_{21} + x_{22} = 1, \tag{9.17}$$

$$x_{ij} \geq 0 \quad i \in \{1, 2\}, \ j \in \{1, 2\},$$

$$u_1 \geq \frac{1}{5},$$

$$u_2 \geq \frac{1}{5}.$$

We use Mathematica to construct a solution with the following code.

```
Maximize[
    {
        (u1-1/5)*(u2-1/5), (*Objective function*)
        2x11 - x12 - x21 + x22 - u1 == 0,
            (*Constraint 1*)
        x11 - x12 - x21 + 2x22 - u2 == 0,
            (*Constraint 2*)
        x11 + x12 + x21 + x22 == 1, (*Constraint 3*)
        x11 >= 0, x12 >= 0, x21 >= 0, x22 >= 0,
        u1 >= 1/5, u2 >= 1/5
    },
    {x11,x12,x21,x22,u1,u2} (*Variables*)
]
```

The solution is

```
{169/100, {x11 -> 1/2, x12 -> 0, x21 -> 0,
            x22 -> 1/2, u1 -> 3/2, u2 -> 3/2}}.
```

Note that we have $u_1 = u_2 = \frac{3}{2}$, as required by symmetry. Interpreting the solution, this means that Players 1 and 2 should flip a fair coin to decide whether they will both follow strategy 1 or strategy 2 (i.e., boxing or ballet). This solution is shown on the set $P(\mathbf{A}, \mathbf{B})$ in Fig. 9.3. Note that the resulting solution is now on the Pareto frontier, as expected.

Example 9.40. It is interesting to consider what happens when bargaining begins from one of the other Nash equilibrium points. Both of those points are already Pareto optimal. (See Fig. 9.2). Consequently, starting at, say, $x_1 = 1$, $x_2 = 0$, $y_1 = 1$, and $y_2 = 0$, we see that there is no way to ensure that each player receives at least as much under the bargaining strategy as they do under the Nash equilibrium while moving to a fairer position. Thus, the Nash bargaining solution when starting at $x_1 = 1$, $x_2 = 0$, $y_1 = 1$, and $y_2 = 0$ is again $x_1 = 1$, $x_2 = 0$, $y_1 = 1$, and $y_2 = 0$. This illustrates a known principle: Always negotiate from a position of strength. In this example, Player 1 is already getting what (s)he wants and has no incentive to negotiate anything away.

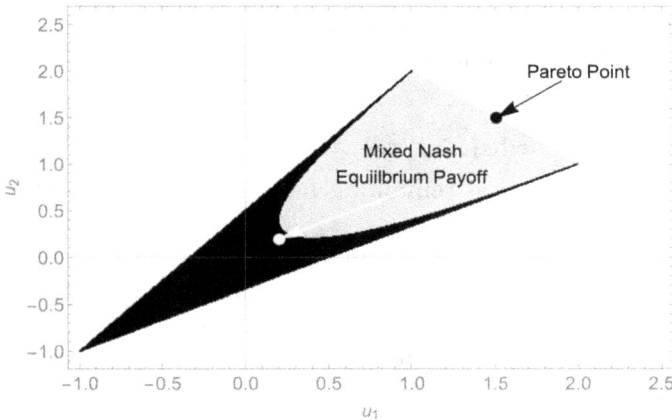

Fig. 9.3. The Pareto-optimal Nash bargaining solution to the Battle of the Sexes game is for each player to do what makes them happiest 50% of the time. This seems like the basis for a fairly happy marriage, and it yields a Pareto-optimal solution, shown by the green dot.

9.5   Chapter Notes

Nash presented this approach to bargaining in his 1953 paper on the topic [110]. It is straightforward to generalize these results to games with $N > 2$ players. The objective function of the Nash bargaining quadratic programming problem is modified to obtain a product of more than two terms. Nash bargaining is a subtopic of the broader field of cooperative bargaining or bargaining theory [111]. Because of its algorithmic nature, Nash bargaining has found several uses in computer science, especially in resource allocation and network design (see, e.g., Refs. [112–115]).

Multi-criteria optimization is also called multi-objective optimization and occurs in a wide range of real-world settings. The simple act of buying a car, where price, fuel efficiency, acceleration, cargo capacity, and visual appeal, is an exercise in multi-criteria optimization. As already discussed, there are several ways of transforming this vector optimization problem into a single-objective optimization problem. A method not already mentioned is sometimes called an ϵ-constraint method [116], though in more restricted settings, it can be called goal programming. In this setting, a single-objective function is chosen for optimization, and the remaining objective functions are given minimum (or maximum) targets and used in additional constraints. These problems then have the form

$$\max z_k(x_1, \ldots, x_n),$$

$$\text{s.t. } z_j(x_1, \ldots, x_n) \geq \epsilon_j \quad j \neq k \quad \mathbf{x} \in X,$$

where X represents the remaining feasible region from the original problem.

Because solutions to these problems usually become philosophical (i.e., which objective is the most important?), it is sometimes more helpful to quantify the Pareto frontier. This can be accomplished by varying the objective function weights when using a linear combination of the objective functions to make a single objective. That is,

$$z(\mathbf{x}; \boldsymbol{\lambda}) = \sum_{i=1}^{s} \lambda_i z_i(\mathbf{x}).$$

Here, $\boldsymbol{\lambda} = \langle \lambda_1, \ldots, \lambda_s \rangle$ are the weights. Varying $\boldsymbol{\lambda}$ and solving the single-criterion optimization problem with the objective function $z(\mathbf{x}; \boldsymbol{\lambda})$ will define the Pareto frontier [116]. This process can be time-consuming, and recent approaches attempt to use deep learning [117] to estimate the Pareto frontier from a few examples.

$$- \spadesuit \clubsuit \heartsuit \diamondsuit -$$

9.6 Exercises

9.1 Prove Lemma 9.8. [Hint: Argue that any pair of mixed strategies can be used to generate a cooperative mixed strategy.]

9.2 Find a Nash equilibrium for the Battle of the Sexes game using a quadratic programming problem.

9.3 Use Nash's bargaining theorem to show that players should trust each other and cooperate rather than defect in prisoner's dilemma.

9.4 Solve the Nash bargaining problem for the Chicken game for each of its three Nash equilibria.

Chapter 10

An Introduction to N-Player Cooperative Games

Chapter Goals: In this final chapter, we introduce some elementary results about N-player cooperative games, which extend the work we began in the previous chapter on bargaining games. We continue to assume that the players in this game can communicate with each other. This subject is a bit different from what has been presented in the previous chapters. Here, the goal is to study games in which it is in the players' best interest to come together in a *grand coalition* of cooperating players. Thus, we define what a coalition is and how payoffs may be assigned within groups of players. We introduce the concept of essential games and the core. We also use our work from linear programming to discuss the Bondareva–Shapley theorem as well as Shapley values.

10.1 Motivating Cooperative Games

Definition 10.1 (Coalition of Players). Consider an N-player game $\mathcal{G} = (\mathbf{P}, \Sigma, \pi)$. Any set $S \subseteq \mathbf{P}$ is called a *coalition of players*. The set $S^c = \mathbf{P} \backslash S$ is the *dual coalition*. The coalition $\mathbf{P}$ itself is called the *grand coalition*.

Remark 10.2. Heretofore, we have always written $\mathbf{P} = \{P_1, \ldots, P_N\}$; however, for the remainder of the chapter, we assume that $\mathbf{P} = \{1, \ldots, N\}$. This will substantially simplify our notation.

Remark 10.3 (Two Coalition Games). Let $\mathcal{G} = (\mathbf{P}, \Sigma, \pi)$ be an N-player game. Suppose within a coalition $S \subseteq \mathbf{P}$ with $S = \{i_1, \ldots, i_{|S|}\}$, the players $i_1, \ldots, i_{|S|}$ agree to play some strategy,

$$\boldsymbol{\sigma}_S = (\sigma_{i_{i_1}}, \ldots, \sigma_{i_{|S|}}) \in \Sigma_1 \times \cdots \times \Sigma_{i_{|S|}},$$

while players in $S^c = \{j_1, \ldots, j_{|S^c|}\}$ agree to play the strategy

$$\boldsymbol{\sigma}_{S^c} = (\sigma_{j_1}, \ldots, \sigma_{j_{|S^c|}}).$$

Under these assumptions, we may suppose that the net payoff to coalition S is

$$K_S = \sum_{i \in S} \pi_i(\boldsymbol{\sigma}_S, \boldsymbol{\sigma}_{S^c}). \tag{10.1}$$

That is, the cumulative payoff to coalition S is simply the sum of the payoffs of the members of the coalition from the payoff function π in the game $\mathcal{G}$. The payoff to the players in S^c is defined similarly as K_{S^c}. Then, we can think of the coalitions as playing a two-player general-sum game, with payoff functions given by K_S and K_{S^c}.

Definition 10.4 (Two-Coalition Game). Given an N-player game $\mathcal{G} = (\mathbf{P}, \Sigma, \pi)$ and a coalition $S \subseteq \mathbf{P}$, with $S = \{i_1, \ldots, i_{|S|}\}$ and $S^c = \{j_1, \ldots, j_{|S^c|}\}$. The two-coalition game is the two-player game

$$\mathcal{G}_S = \Big(\{S, S^c\}, \Big(\Sigma_{i_1} \times \cdots \times \Sigma_{i_{|S|}}\Big) \times \Big(\Sigma_{j_1} \times \cdots \times \Sigma_{j_{|S^c|}}\Big),$$

$$(K_S \times K_{S^c})\Big).$$

Remark 10.5. We leave the proof of the following lemma as an exercise.

Lemma 10.6. *For any two-coalition game $\mathcal{G}_S$, there is a Nash equilibrium strategy for both the coalition S and its dual S^c.*

Definition 10.7 (Characteristic (Value) Function). Let S be a coalition defined over an N-player game $\mathcal{G}$. Then, the value function $v \colon 2^{\mathbf{P}} \to \mathbb{R}$ is the expected payoff to S in the game $\mathcal{G}_S$ when both coalitions S and S^c play their Nash equilibrium strategy in $\mathcal{G}_S$.

Remark 10.8. The characteristic or value function can be considered the net worth of the coalition to its members. Clearly,

$$v(\emptyset) = 0$$

because the empty coalition can achieve no value. On the other hand,

$$v(\mathbf{P}) = \text{the largest sum of all payoff values possible}$$

because a two-player game against the empty coalition will try to maximize the value of Eq. (10.1). In general, $v(\mathbf{P})$ answers the following question: If all N players worked together to maximize the sum of their payoffs, which strategy would they all agree to choose, and what would that sum be?

Theorem 10.9. *If* $S, T \subseteq \mathbf{P}$ *and* $S \cap T = \emptyset$, *then* $v(S) + v(T) \leq v(S \cup T)$.

Proof. Within S and T, the players may choose a strategy (jointly) and independently to ensure that they receive at least $v(S) + v(T)$; however, the value of the game $\mathcal{G}_{S \cup T}$ to Player 1 $(S \cup T)$ may be larger than the result yielded when S and T make independent choices, thus $v(S \cup T) \geq v(S) + v(T)$. $\square$

Definition 10.10 (Superadditivity). A function $v: 2^{\mathbf{P}} \to \mathbb{R}$ is *superadditive* if

(1) $v(\emptyset) = 0$;
(2) $v(S) + v(T) \leq v(S \cup T)$, for all $S, T \subseteq \mathbf{P}$, assuming $S \cap T = \emptyset$.

Remark 10.11. Theorem 10.9 states that when the function $v: 2^{\mathbf{P}} \to \mathbb{R}$ is defined in terms of the Nash equilibrium payoff in the two-coalition game, then the resulting function is superadditive.

10.2   Coalition Games

Remark 10.12. The goal of cooperative N-player games is to define scenarios in which the *grand coalition*, $\mathbf{P}$, is stable; that is, it is in everyone's interest to work together in one large coalition, $\mathbf{P}$. The problem then becomes to divide the coalition payoff $v(S)$ among the members of the coalition S, and by being in a coalition, the players will improve their payoff over competing on their own.

Definition 10.13 (Coalition Game). A coalition game is a pair $(\mathbf{P}, v)$ where $\mathbf{P}$ is the set of players and $v\colon 2^{\mathbf{P}} \to \mathbb{R}$ is a superadditive characteristic function.

Remark 10.14. At this point, v can be any superadditive function. It is not necessarily defined by the two-coalition game.

Definition 10.15 (Inessential Game). A game is *inessential* if

$$v(\mathbf{P}) = \sum_{i=1}^{N} v(\{i\}).$$

A game that is not inessential is called *essential*.

Remark 10.16. An inessential game is one in which the total value of the grand coalition does *not* exceed the sum of the values to the players if they each played against all other players individually. That is, there is no incentive for any player to join the grand coalition because there is no chance that they will receive a better payoff, assuming the total payoff of the grand coalition is divided among its members.

Theorem 10.17. *Let $S \subseteq \mathbf{P}$. In an inessential game,*

$$v(S) = \sum_{i \in S} v(\{i\}).$$

Proof.   We proceed by contradiction. Suppose not; then,

$$v(S) > \sum_{i \in S} v(\{i\})$$

by superadditivity. Now,

$$v(S^c) \geq \sum_{i \in S^c} v(\{i\}),$$

and $v(\mathbf{P}) \geq v(S) + v(S^c)$, which implies that

$$v(\mathbf{P}) \geq v(S) + v(S^c) > \sum_{i \in S} v(\{i\}) + \sum_{i \in S^c} v(\{i\}) = \sum_{i \in \mathbf{P}} v(\{i\}).$$

Thus,

$$v(\mathbf{P}) > \sum_{i=1}^{N} v(\{i\});$$

therefore, the coalition game is not inessential. This is a direct contradiction of our assumption. □

Corollary 10.18. *A two-player zero-sum game produces an inessential coalition game.*

10.3   Division of Payoff to the Coalition

Remark 10.19. Given a coalition game, $(\mathbf{P}, v)$, the goal is to equitably divide $v(S)$ among the members of the coalition in such a way that the individual players prefer to be in the coalition rather than to leave it.

Definition 10.20 (Imputation). Let $(\mathbf{P}, v)$ be a coalition game. A tuple $(x_1, \ldots, x_N)$ (of payoffs to the individual players in $\mathbf{P}$) is called an *imputation* if:

(1) $x_i \geq v(\{i\})$ and
(2) $\sum_{i \in \mathbf{P}} x_i = v(\mathbf{P})$.

Remark 10.21. The first criterion for a tuple $(x_1, \ldots, x_N)$ to be an imputation states that each player must do better in the grand coalition than they would on their own (against all other players). The second criterion says that the total allotment of payoff to the players cannot exceed the payoff received by the grand coalition itself. Essentially, this second criterion asserts that the coalition cannot go into debt to maintain its members. It is also worth noting that the condition

$$\sum_{i \in \mathbf{P}} x_i = v(\mathbf{P})$$

is equivalent to a statement on Pareto optimality in so far as players all together cannot expect to do any better than the net payoff accorded to the grand coalition.

Definition 10.22 (Dominance). Let $(\mathbf{P}, v)$ be a coalition game. Suppose $\mathbf{x} = (x_1, \ldots, x_N)$ and $\mathbf{y} = (y_1, \ldots, y_N)$ are two imputations. Then, $\mathbf{x}$ *dominates* $\mathbf{y}$ over some coalition $S \subset \mathbf{P}$ if:

(1) $x_i > y_i$, for all $i \in S$, and
(2) $\sum_{i \in S} x_i \leq v(S)$.

Remark 10.23. The previous definition states that players in coalition S prefer the payoffs they receive under $\mathbf{x}$ to the payoffs they receive under $\mathbf{y}$. Furthermore, these same players can threaten to *leave the grand coalition* $\mathbf{P}$ because they may actually improve their payoff by playing coalition S.

Definition 10.24 (Stable Set). A stable set $X \subseteq \mathbb{R}^n$ of imputations is a set satisfying the following criteria:

(1) No payoff vector $\mathbf{x} \in X$ is dominated over any coalition by another imputation, $\mathbf{y} \in X$.
(2) All payoff vectors $\mathbf{y} \notin X$ are dominated by at least one vector, $\mathbf{x} \in X$.

Remark 10.25. Stable sets are (in some way) good sets of imputations in so far as they represent imputations that will make players want to remain in the grand coalition.

10.4   The Core

Definition 10.26 (Core). Given a coalition game $(\mathbf{P}, v)$, the core is the set of imputations

$$C(v) = \left\{ \mathbf{x} \in \mathbb{R}^n \colon \sum_{i=1}^N x_i = v(\mathbf{P}) \text{ and } \forall S \subseteq \mathbf{P} \left(\sum_{i \in S} x_i \geq v(S) \right) \right\}.$$

Remark 10.27. A vector $\mathbf{x} \in C(v)$ is an imputation. This is clear, as

$$\sum_{i \in \mathbf{P}} x_i = v(\mathbf{P}),$$

and since $\{i\} \subset \mathbf{P}$, we know that $x_i \geq v(\{i\})$. However, the core (when it is non-empty) is more than that.

Theorem 10.28. *The core is contained in every stable set.*

Proof.   Let X be a stable set. If the core is empty, then it is contained in X. Therefore, suppose $\mathbf{x} \in C(v)$. If $\mathbf{x}$ is dominated by any vector $\mathbf{z}$, then there is a coalition $S \subset \mathbf{P}$ so that $z_i > x_i$ for all $i \in S$

and $\sum_{i \in S} z_i \leq v(S)$. However, by the definition of the core, we know that

$$\sum_{i \in S} z_i > \sum_{i \in S} x_i \geq v(S).$$

Thus, $\sum_{i \in S} z_i > v(S)$ and $\mathbf{z}$ cannot dominate $\mathbf{x}$ – a contradiction.

□

Remark 10.29. We can use linear programming and the definition of the core to develop a test for the emptiness of the core. The following theorem follows immediately from the definition of the core.

Theorem 10.30. *Let* $(\mathbf{P}, v)$ *be a coalition game. Consider the linear programming problem*

$$\begin{cases} \min \ x_1 + \cdots + x_N, \\ s.t. \ \sum_{i \in S} x_i \geq v(S) \quad for \ all \ S \subseteq \mathbf{P}. \end{cases} \tag{10.2}$$

If there is no solution $\mathbf{x}^*$ *so that* $\sum_{i=1}^{N} x_i = v(P)$, *then* $C(v) = \emptyset$.

Remark 10.31. We minimize the sum in the objective since we want to find small enough values for $x_1, \ldots, x$ so that we can form an imputation.

Corollary 10.32. *The core of a coalition game,* $(\mathbf{P}, v)$, *may be empty.*

Remark 10.33. We can now use linear programming duality to prove the Bondareva–Shapley Theorem. We state a lemma to help compute a dual linear programming problem. The lemma can be proved by showing that the two problems share the Karush–Kuhn–Tucker (KKT) conditions, with primal feasibility and dual feasibility swapped.

Lemma 10.34. *Suppose* $\mathbf{A} \in \mathbb{R}^{m \times n}$, $\mathbf{c} \in \mathbb{R}^n$, $\mathbf{x} \in \mathbb{R}^n$ *is a set of variables, and* $\mathbf{y} \in \mathbb{R}^m$ *is a set of dual variables (Lagrange multipliers). The linear programming problem*

$$\begin{cases} \min \ \mathbf{c}^T \mathbf{x}, \\ s.t. \ \mathbf{A}\mathbf{x} \geq \mathbf{b}, \\ \mathbf{x} \ unrestricted \end{cases}$$

has the dual problem

$$
\begin{cases}
\max \ \mathbf{b}^T\mathbf{y}, \\
s.t. \ \mathbf{A}^T\mathbf{y} = \mathbf{c}, \\
\mathbf{y} \geq \mathbf{0}.
\end{cases}
$$

Theorem 10.35 (Bondareva–Shapley Theorem). *Let $(\mathbf{P}, v)$ be a coalition game with $|\mathbf{P}| = N$. The core $C(v)$ is non-empty if and only if there exists $y_1, \ldots, y_{2^N}$, where each y_i corresponds to a set $S_i \subseteq \mathbf{P}$ so that*

$$
\begin{cases}
v(\mathbf{P}) = \displaystyle\sum_{i=1}^{2^N} y_i v(S_i), \\
\displaystyle\sum_{S_i \supseteq \{j\}} y_i = 1 \quad \forall j \in \mathbf{P}, \\
y_i \geq 0 \quad \forall S_i \subseteq \mathbf{P}.
\end{cases}
$$

Proof. Apply Lemma 10.34 to compute the dual linear programming problem for Eq. (10.2) as

$$
\begin{cases}
\max \ \displaystyle\sum_{i=1}^{2^N} y_i v(S_i), \\
s.t. \ \displaystyle\sum_{S_i \supseteq \{j\}} y_i = 1 \quad \forall j \in \mathbf{P}, \\
y_i \geq 0 \quad \forall S_i \subseteq \mathbf{P}.
\end{cases}
\tag{10.3}
$$

To see this, we note that there are 2^N constraints in Eq. (10.2) and N variables. Thus, there will be N constraints in the dual problem, but 2^N variables (Lagrange multipliers), and the resulting dual problem can be written as Eq. (10.3). By Theorem 7.29 (the strong duality theorem), Eq. (10.3) has a solution if and only if Eq. (10.2) does; moreover, the objective functions at optimality coincide. This completes the proof. □

Corollary 10.36. *A non-empty core is not necessarily a singleton.*

Remark 10.37. The core can be considered the possible "equilibrium" imputations that smart players will agree to and that cause the grand coalition to hold together; i.e., no players or group of players have any motivation to leave the grand coalition. Unfortunately, the fact that the core may be empty is not helpful.

10.5 Shapley Values

Definition 10.38 (Shapley Values). Let $(\mathbf{P}, v)$ be a coalition game with N players. Then, the Shapley value for Player i is

$$x_i = \phi_i(v) = \sum_{S \subseteq \mathbf{P} \backslash \{i\}} \frac{|S|!(N - |S| - 1)!}{N!} (v(S \cup \{i\}) - v(S)). \quad (10.4)$$

Remark 10.39. The Shapley value is the *average extra value* Player i contributes to each possible coalition that might form. Imagine forming the grand coalition one player at a time. There are $N!$ ways to do this. Hence, in an average, $N!$ is in the denominator of the Shapley value.

Now, if we have formed a coalition S (on our way to forming $\mathbf{P}$), then there are $|S|!$ ways we could have done this. Each of these ways yields $v(S)$ in value because the characteristic function does not value how a coalition is formed, only the members of the coalition.

Once we add Player i to the coalition S, the new value is $v(S \cup \{i\})$, and the value Player i added was $v(S \cup \{i\}) - v(S)$. We then add the other $N - |S| - 1$ players to achieve the grand coalition. There are $(N - |S| - 1)!$ ways of doing this.

Thus, the extra value Player i adds in each case is $v(S \cup \{i\}) - v(S)$ multiplied by $|S|!(N - |S| - 1)!$ for each of the possible ways this exact scenario occurs. Summing over all possible subsets S and dividing by $N!$, as noted, yields the average excess value Player i brings to a coalition.

Remark 10.40. We state, but do not prove, the following theorem. The proof rests on the linear properties of averages. That is, we note that Eq. (10.4) is a linear expression in $v(S)$ and $v(S \cup \{i\})$.

Theorem 10.41. *For any coalition game* $(\mathbf{P}, v)$ *with* N *players:*

(1) $\phi_i(v) \geq v(\{i\})$.

(2) $\sum_{i \in \mathbf{P}} \phi_i(v) = v(\mathbf{P})$.

(3) *From* (1) *and* (2), *we conclude that* $(\phi_1(v), \ldots, \phi_N(v))$ *is an imputation.*

(4) *If for all* $S \subseteq \mathbf{P}$, $v(S \cup \{i\}) = v(S \cup \{j\})$ *with* $i, j \notin S$, *then* $\phi_i(v) = \phi_j(v)$.

(5) *If* v *and* w *are two characteristic functions in coalition games* $(\mathbf{P}, v)$ *and* $(\mathbf{P}, w)$, *then* $\phi_i(v + w) = \phi_i(v) + \phi_i(w)$ *for all* $i \in \mathbf{P}$.

(6) *If* $v(S \cup \{i\}) = v(S)$ *for all* $S \subseteq \mathbf{P}$ *with* $i \notin S$, *then* $\phi_i(v) = 0$ *because Player* i *contributes nothing to the grand coalition.*

Example 10.42 (Extended Example). Consider the coalition game $(\mathbf{P}, v)$, with $\mathbf{P} = \{1, 2, 3\}$, and v defined by

$$v(\{1, 2, 3\}) = 6,$$
$$v(\{1, 2\}) = 2,$$
$$v(\{1, 3\}) = 6,$$
$$v(\{2, 3\}) = 4,$$
$$v(\{1\}) = v(\{2\}) = v(\{3\}) = 0.$$

We can see at once that this game is essential because

$$v(\mathbf{P}) \neq v(\{1\}) + v(\{2\}) + v(\{3\}).$$

We can also write the linear programming problem to identify the core or determine whether it is empty. There are seven sets to consider: $\{1\}$, $\{2\}$, $\{3\}$, $\{1, 2\}$, $\{1, 3\}$, $\{2, 3\}$, and $\{1, 2, 3\}$. We can ignore the empty set. This gives the linear programming problem

$$\min \ x_1 + x_2 + x_3,$$
$$s.t. \ x_1 \geq 0,$$
$$x_2 \geq 0,$$
$$x_3 \geq 0,$$
$$x_1 + x_2 \geq 2,$$
$$x_1 + x_3 \geq 6,$$

$$x_2 + x_3 \geq 4,$$

$$x_1 + x_2 + x_3 \geq 6.$$

A solution to this linear programming problem is $x_1 = 2$, $x_2 = 0$, and $x_3 = 4$. Thus, the core is not empty.

Lastly, we can find the Shapley values for this game. First, consider the set $\mathbf{P}\backslash\{1\} = \{2, 3\}$. The subsets are: (i) $S = \{2\}$, with $|S|! = 1$ and $(N - |S| - 1)! = 1$; (ii) $S = \{3\}$, with $|S|! = 1$ and $(N - |S| - 1)! = 1$; and (iii) $S = \{2, 3\}$, with $|S|! = 2$ and $(N - |S| - 1)! = 1$. Using the formula for Shapley values, we have

$$x_1 = \frac{1}{6}(2 - 0) + \frac{1}{6}(6 - 0) + \frac{2}{6}(6 - 4) = 2.$$

Now, consider the set $\mathbf{P}\backslash\{2\} = \{1, 3\}$. Using the same approach as above, we have

$$x_2 = \frac{1}{6}(2 - 0) + \frac{1}{6}(4 - 0) + \frac{2}{6}(6 - 6) = 1.$$

Finally, consider the set $\mathbf{P}\backslash\{3\} = \{1, 2\}$. Then, we have

$$x_3 = \frac{1}{6}(6 - 0) + \frac{1}{6}(4 - 0) + \frac{2}{6}(6 - 2) = 3.$$

A quick check shows that $x_1 + x_2 + x_3 = 6$, as required. You will also notice, by plugging these values into the linear programming constraints, that this is not an alternative optimal solution to the linear programming problem. This shows that the Shapley values do not have to produce an imputation in the core.

10.6   Chapter Notes

Cooperative games and coalition games are of substantial interest in economics. Luce and Raiffa [1] and Myerson [38] go into significantly more detail on the subject. This chapter provides only a basic introduction to an exceptionally broad field.

The stable set approach was originally developed by von Neumann and Morgenstern [5]. The core was not considered by von Neumann and Morgenstern because their text focused on zero-sum games, whereas we have shown that such games are inessential and consequently the core is always empty. Donald Gillies, a Canadian mathematician and computer scientist, introduced the idea of the core in 1959 [118]. Shapley values were introduced by Lloyd Shapley in 1953 [119]. Shapley worked in several areas of game theory, including stochastic games, cooperative games, and the stable marriage problem, and, with Alvin Roth, won the 2012 Nobel Memorial Prize in Economic Sciences for his contributions to the subject.

$$- \spadesuit \clubsuit \heartsuit \diamondsuit -$$

10.7 Exercises

10.1 Prove Lemma 10.6. [Hint: Use Nash's theorem.]

10.2 Prove Corollary 10.18.

10.3 Consider the coalition game (P, v), with $P = \{1, 2, 3\}$, and v defined by

$$v(\{1, 2, 3\}) = 10,$$
$$v(\{1, 2\}) = 4,$$
$$v(\{1, 3\}) = 8,$$
$$v(\{2, 3\}) = 6,$$
$$v(\{1\}) = v(\{2\}) = v(\{3\}) = 0.$$

(a) Is this game essential or inessential? Why?
(b) Write a linear programming problem that will determine whether the core is empty. (You do not need to solve this linear programming problem.)
(c) Find the Shapely value for each player.

10.4 Use a linear programming solver to find an imputation in the core of the game in Exercise 10.3 or determine whether the core is empty.

10.5 Explicitly construct the dual linear programming problem in Theorem 10.35 and fill in the details of the proof using Lemma 10.34.

10.6   Prove Corollary 10.36. [Hint: Think about alternative optimal solutions.]

10.7   Show that computing the core is an exponential problem, even though solving a linear programming problem is known to be polynomial in the size of the problem. [Hint: How many constraints are there?]

10.8   Prove Theorem 10.41.

Appendix A

Introduction to Matrix Arithmetic

Appendix Goals: This appendix introduces the essentials of matrix arithmetic needed for studying matrix games. Proofs are omitted for brevity. The goals of this appendix are to introduce matrices and vectors in $\mathbb{R}^n$ as well as special matrices, such as the identity matrix. Matrix operations like addition, multiplication, and transpose are also discussed.

A.1 Matrices, Row and Column Vectors

Definition A.1 (Matrix). An $m \times n$ matrix is a rectangular array of values (*scalars*), drawn from $\mathbb{R}$. We write $\mathbb{R}^{m \times n}$ to denote the set of $m \times n$ matrices with entries drawn from $\mathbb{R}$.

Example A.2. Here is an example of a 2×3 matrix drawn from $\mathbb{R}^{2 \times 3}$:

$$\mathbf{A} = \begin{bmatrix} 3 & 1 & \frac{7}{2} \\ 2 & \sqrt{2} & 5 \end{bmatrix}.$$

Remark A.3. We denote the element at position (i, j) of matrix $\mathbf{A}$ as $\mathbf{A}_{ij}$. Thus, in the example above, $\mathbf{A}_{2,1} = 2$.

Remark A.4. We restrict our attention to matrices with real entries. In general, a matrix can take entries from any field (such as the complex numbers). See Ref. [73] for details.

Definition A.5 (Matrix Addition). If $\mathbf{A}$ and $\mathbf{B}$ are both in $\mathbb{R}^{m \times n}$, then $\mathbf{C} = \mathbf{A} + \mathbf{B}$ is the matrix sum of $\mathbf{A}$ and $\mathbf{B}$ in $\mathbb{R}^{m \times n}$, and

$$\mathbf{C}_{ij} = \mathbf{A}_{ij} + \mathbf{B}_{ij} \text{ for } i = 1, \ldots, m \text{ and } j = 1, \ldots, n. \quad (A.1)$$

Here, $+$ is the standard operation for addition.

Example A.6.

$$\begin{bmatrix} 1 & 2 \\ 3 & 4 \end{bmatrix} + \begin{bmatrix} 5 & 6 \\ 7 & 8 \end{bmatrix} = \begin{bmatrix} 1+5 & 2+6 \\ 3+7 & 4+8 \end{bmatrix} = \begin{bmatrix} 6 & 8 \\ 10 & 12 \end{bmatrix}. \quad (A.2)$$

Definition A.7 (Scalar-Matrix Multiplication). If $\mathbf{A}$ is a matrix from $\mathbb{R}^{m \times n}$ and $c \in \mathbb{R}$, then $\mathbf{B} = c\mathbf{A} = \mathbf{A}c$ is the scalar-matrix product of c and $\mathbf{A}$ in $\mathbb{R}^{m \times n}$, and

$$\mathbf{B}_{ij} = c\mathbf{A}_{ij} \text{ for } i = 1, \ldots, m \text{ and } j = 1, \ldots, n. \quad (A.3)$$

Example A.8. Let

$$\mathbf{B} = \begin{bmatrix} 3 & 7 \\ 6 & 3 \end{bmatrix}.$$

When we multiply by the scalar $3 \in \mathbb{R}$, we obtain

$$3 \cdot \begin{bmatrix} 3 & 7 \\ 6 & 3 \end{bmatrix} = \begin{bmatrix} 9 & 21 \\ 18 & 9 \end{bmatrix}.$$

Definition A.9 (Row/Column Vector). A $1 \times n$ matrix is called a *row vector*, and a $m \times 1$ matrix is called a *column vector*. Unless specified otherwise, every vector is considered a **column vector**. A column vector $\mathbf{x}$ in $\mathbb{R}^{n \times 1}$ (or $\mathbb{R}^n$) is $\mathbf{x} = \langle x_1, \ldots, x_n \rangle$.

Remark A.10. It should be clear that any row of the matrix $\mathbf{A}$ could be considered a row vector, and any column of $\mathbf{A}$ could be considered a column vector. The ith row of $\mathbf{A}$ is denoted $\mathbf{A}_{i\cdot}$, while the jth column is denoted $\mathbf{A}_{\cdot j}$. Also, any row/column vector is nothing more sophisticated than tuples of numbers (a point in space). You are free to think of these things however you like.

A.2 Matrix Multiplication

Definition A.11 (Dot Product). If $\mathbf{x}, \mathbf{y} \in \mathbb{R}^n$ are two n-dimensional vectors, then the *dot product* is

$$\mathbf{x} \cdot \mathbf{y} = \sum_{i=1}^{n} x_i y_i, \tag{A.4}$$

where x_i is the ith component of the vector $\mathbf{x}$.

Remark A.12. The dot product is an example of a more general concept called an *inner product*, which maps two vectors to a scalar. Not all inner products behave according to Eq. (A.4), and the definition of the inner product will be context dependent.

Definition A.13 (Matrix Multiplication). If $\mathbf{A} \in \mathbb{R}^{m \times n}$ and $B \in \mathbb{R}^{n \times p}$, then $\mathbf{C} = \mathbf{AB}$ is the *matrix product* of $\mathbf{A}$ and $\mathbf{B}$, and

$$\mathbf{C}_{ij} = \mathbf{A}_{i\cdot} \cdot \mathbf{B}_{\cdot j}. \tag{A.5}$$

Note that $\mathbf{A}_{i\cdot} \in \mathbb{R}^{1 \times n}$ (an n-dimensional vector) and $\mathbf{B}_{\cdot j} \in \mathbb{R}^{n \times 1}$ (another n-dimensional vector), thus making the dot product meaningful. Note also that $\mathbf{C} \in \mathbb{R}^{m \times p}$.

Example A.14.

$$\begin{bmatrix} 1 & 2 \\ 3 & 4 \end{bmatrix} \begin{bmatrix} 5 & 6 \\ 7 & 8 \end{bmatrix} = \begin{bmatrix} 1(5) + 2(7) & 1(6) + 2(8) \\ 3(5) + 4(7) & 3(6) + 4(8) \end{bmatrix} = \begin{bmatrix} 19 & 22 \\ 43 & 50 \end{bmatrix}. \tag{A.6}$$

Remark A.15. Note that we cannot multiply any pair of arbitrary matrices. If we have the product $\mathbf{AB}$ of two matrices $\mathbf{A}$ and $\mathbf{B}$, then the number of columns in $\mathbf{A}$ must be equal to the number of rows in $\mathbf{B}$.

Definition A.16 (Matrix Transpose). If $\mathbf{A} \in \mathbb{R}^{m \times n}$ is an $m \times n$ matrix, then the *transpose* of $\mathbf{A}$, denoted $\mathbf{A}^T$, is an $m \times n$ matrix defined as

$$\mathbf{A}_{ij}^T = \mathbf{A}_{ji}. \tag{A.7}$$

Example A.17.

$$\begin{bmatrix} 1 & 2 \\ 3 & 4 \end{bmatrix}^T = \begin{bmatrix} 1 & 3 \\ 2 & 4 \end{bmatrix}. \tag{A.8}$$

Essentially, we are just reading down the columns of $\mathbf{A}$ to obtain its transpose.

Remark A.18. The matrix transpose is a particularly useful operation and makes it easy to transform column vectors into row vectors, which enables multiplication. For example, suppose $\mathbf{x}$ is an $n \times 1$ column vector (i.e., $\mathbf{x}$ is a vector in $\mathbb{R}^n$), and suppose $\mathbf{y}$ is an $n \times 1$ column vector. Then,

$$\mathbf{x} \cdot \mathbf{y} = \mathbf{x}^T \mathbf{y}. \tag{A.9}$$

Proposition A.19. *Suppose* $\mathbf{A}, \mathbf{B} \in \mathbb{R}^{n \times n}$. *Then,*

$$(\mathbf{AB})^T = \mathbf{B}^T \mathbf{A}^T.$$

A.3 Special Matrices and Vectors

Remark A.20. There are many special (and useful) square matrices, as we discuss in the following.

Definition A.21 (Standard Basis Vector). The ith *standard basis vector* in $\mathbb{R}^n$ is the vector $\mathbf{e}_i = \langle 0, \ldots, 1, 0, \ldots, 0 \rangle$, where the number 1 occurs at position i.

Definition A.22 (Diagonal Matrix). A *diagonal matrix* is a square matrix with the property that $\mathbf{D}_{ij} = 0$, for $i \neq j$ and where $\mathbf{D}_{ii}$ may take any value in $\mathbb{R}$.

Example A.23. Consider the matrix

$$\mathbf{D} = \begin{bmatrix} 2 & 0 & 0 \\ 0 & 4 & 0 \\ 0 & 0 & 6 \end{bmatrix}.$$

This is a diagonal matrix.

Definition A.24 (Identity Matrix). The $n \times n$ diagonal matrix $\mathbf{I}_n$ with ones along the diagonal is called the *identify matrix*. That is,

$$\mathbf{I}_n = \begin{bmatrix} 1 & \cdots & 0 \\ \vdots & \ddots & \vdots \\ 0 & \cdots & 1 \end{bmatrix}. \tag{A.10}$$

Remark A.25. Note that the identity matrix consists of rows and columns made up of standard basis vectors.

Proposition A.26. *Let* $\mathbf{A} \in \mathbb{R}^{n \times n}$. *Then,*

$$\mathbf{A}\mathbf{I}_n = \mathbf{I}_n\mathbf{A}.$$

Definition A.27 (Zero Vector). The $n \times 1$ *zero* vector, denoted $\mathbf{0}$, is an $n \times 1$ vector consisting of all zeros. The dimension n is determined from context, when possible. If the dimension of the zero vector cannot be determined from context, we denote it $\mathbf{0}_n$.

Definition A.28 (One Vector). The $n \times 1$ *one* vector, denoted $\mathbf{1}$, is an $n \times 1$ vector consisting of all ones. The dimension n is determined from context, when possible. If the dimension of the zero vector cannot be determined from context, we denote it $\mathbf{1}_n$.

Definition A.29 (Symmetric Matrix). Let $\mathbf{M} \in \mathbb{R}^{n \times n}$ be a matrix. The matrix $\mathbf{M}$ is symmetric if $\mathbf{M} = \mathbf{M}^T$.

Example A.30. Suppose that

$$\mathbf{A} = \begin{bmatrix} 1 & 2 & 3 \\ 2 & 4 & 1 \\ 3 & 1 & 7 \end{bmatrix}.$$

This matrix is symmetric.

$$- \spadesuit \clubsuit \heartsuit \diamondsuit -$$

A.4   Exercises

A.1   Show that if two matrices $\mathbf{A}, \mathbf{B} \in \mathbb{R}^{m \times n}$, then

$$\mathbf{A} + \mathbf{B} = \mathbf{B} + \mathbf{A},$$

that is, matrix addition is commutative.

A.2   Show by counterexample that if $\mathbf{A}, \mathbf{B} \in \mathbb{R}^{n \times n}$, then it is not necessarily the case that $\mathbf{AB} = \mathbf{BA}$, thus showing that matrix multiplication is not commutative. [Hint: Almost any pair of square matrices that are chosen at random will not commute when multiplied.]

A.3 Prove Proposition A.26.

A.4 Prove Proposition A.19.

Appendix B

Essential Concepts from Vector Calculus

> **Appendix Goals:** This appendix introduces the essentials of vector calculus needed for optimization and hence game theory. Most proofs are omitted for brevity but can be found in any vector calculus text, including Ref. [72]. The goals of this appendix are to introduce dot products, directional derivatives, gradients, and level sets, along with their properties.

B.1   Geometry for Vector Calculus

Remark B.1. The dot product of two vectors is given in Definition A.11 in Appendix A. We now give an alternate characterization, which is useful for optimization.

Definition B.2. Let $\mathbf{x} = \langle x_1, \ldots, x_n \rangle$ be a vector in $\mathbb{R}^n$. The norm (or length) of $\mathbf{x}$ is

$$\|\mathbf{x}\| = \sqrt{\mathbf{x} \cdot \mathbf{x}} = \sqrt{x_1^2 + x_2^2 + \cdots + x_n^2}.$$

Theorem B.3 (Dot Product). *Let θ be the angle between the vectors $\mathbf{x}$ and $\mathbf{y}$ in $\mathbb{R}^n$. Then, the dot product of $\mathbf{x}$ and $\mathbf{y}$ is*

$$\mathbf{x} \cdot \mathbf{y} = \|\mathbf{x}\| \, \|\mathbf{y}\| \cos(\theta). \tag{B.1}$$

Remark B.4. This fact can be proved using the *law of cosines* from trigonometry. As a result, we have the following short proposition (which is proved as Theorem 1 in Ref. [72]).

Proposition B.5. *Let* $\mathbf{x}, \mathbf{y} \in \mathbb{R}^n$. *Then, the following hold:*

(1) *The angle between* $\mathbf{x}$ *and* $\mathbf{y}$ *is less than* $\pi/2$ *(i.e., acute) if and only if* $\mathbf{x} \cdot \mathbf{y} > 0$.
(2) *The angle between* $\mathbf{x}$ *and* $\mathbf{y}$ *is exactly* $\pi/2$ *(i.e., the vectors are orthogonal) if and only if* $\mathbf{x} \cdot \mathbf{y} = 0$.
(3) *The angle between* $\mathbf{x}$ *and* $\mathbf{y}$ *is greater than* $\pi/2$ *(i.e., obtuse) if and only if* $\mathbf{x} \cdot \mathbf{y} < 0$.

Definition B.6 (Graph). Let $z \colon D \subseteq \mathbb{R}^n \to \mathbb{R}$ be a function. Then, the *graph* of z is the set of $n + 1$ tuples,

$$G(z) = \{(\mathbf{x}, z(\mathbf{x})) \in \mathbb{R}^{n+1} | \mathbf{x} \in D\}. \tag{B.2}$$

Remark B.7. When $z \colon D \subseteq \mathbb{R} \to \mathbb{R}$, the graph is precisely what you'd expect. It is the set of pairs, $(x, y) \in \mathbb{R}^2$, so that $y = z(x)$. This is the "graph" from a first course in elementary algebra.

Remark B.8. It is inconvenient that the set $G(z)$ is called a *graph*, as this term is also used to refer to the completely distinct discrete structure (which is sometimes also called a *network*) studied in the subject of *graph theory* [39]. Fortunately, these two topics are rarely covered in the same class.

Definition B.9 (Level Set). Let $z \colon \mathbb{R}^n \to \mathbb{R}$ be a function, and let $c \in \mathbb{R}$. Then, the *level set with value c for function z* is the set

$$L_c(z) = \{\mathbf{x} \in \mathbb{R}^n | z(\mathbf{x}) = c\} \subseteq \mathbb{R}^n. \tag{B.3}$$

Example B.10. Consider the function $z(x, y) = x^2 + y^2$. The level set of z at $c = 4$ is the set of points $(x, y) \in \mathbb{R}^2$ such that

$$x^2 + y^2 = 4. \tag{B.4}$$

This is the equation for a circle with a radius of 2. We illustrate this in two ways in Fig. B.1. Figure B.1 (left) shows the level sets of z as they sit on the 3D plot of the function, while Fig. B.1 (right) shows the level sets of z in $\mathbb{R}^2$. The plot in Fig. B.1 (right) is called a *contour plot*. It is important to remember that if $z \colon \mathbb{R}^n \to \mathbb{R}$, then its level sets exist in $\mathbb{R}^n$, while the graph exists in $\mathbb{R}^{n+1}$.

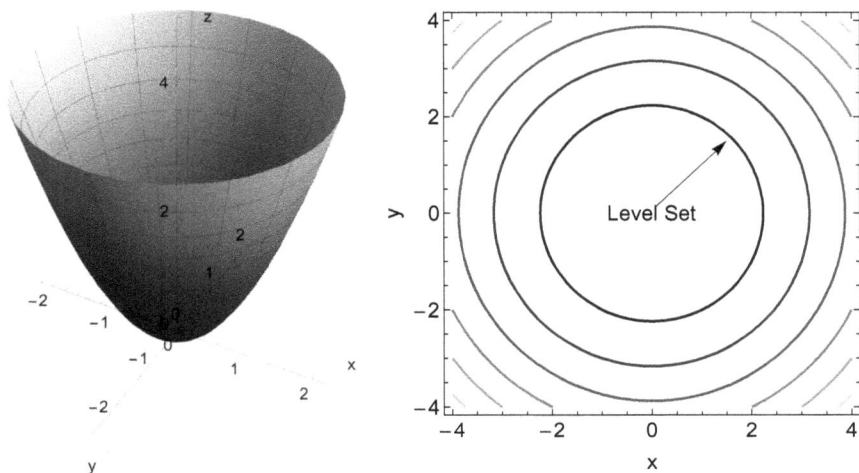

Fig. B.1. (Left) Plot with level sets projected on the graph of z. The level sets exist in $\mathbb{R}^2$, while the graph of z exists in $\mathbb{R}^3$. (Right) Contour Plot of $z = x^2 + y^2$. The circles in $\mathbb{R}^2$ are the level sets of the function.

Definition B.11 (Line). Let $\mathbf{x}_0, \mathbf{v} \in \mathbb{R}^n$. The *line* defined by vectors $\mathbf{x}_0$ and $\mathbf{v}$ is the parametric function $\mathbf{l}(t) = \mathbf{x}_0 + t\mathbf{v}$. The vector $\mathbf{v}$ is called the direction of the line, and $l \colon \mathbb{R} \to \mathbb{R}^n$.

Example B.12. Let $\mathbf{x}_0 = (2, 1)$ and $\mathbf{v} = (2, 2)$. Then, the line defined by $\mathbf{x}_0$ and $\mathbf{v}$ is shown in Fig. B.2. The set of points on this line is the set

$$C = \{(x, y) \in \mathbb{R}^2 \colon x = 2 + 2t, y = 1 + 2t, t \in \mathbb{R}\}.$$

Definition B.13 (Directional Derivative). Let $z \colon \mathbb{R}^n \to \mathbb{R}$, and let $\mathbf{v} \in \mathbb{R}^n$ be a vector (direction). Then, the directional derivative of z at a point $\mathbf{x}_0 \in \mathbb{R}^n$ in the direction of $\mathbf{v}$ is

$$\frac{d}{dt}z(\mathbf{x}_0 + t\mathbf{v})\bigg|_{t=0}, \tag{B.5}$$

when this derivative exists.

Proposition B.14. *The directional derivative of z at $\mathbf{x}_0$ in the direction $\mathbf{v}$ is equal to*

$$\lim_{h \to 0} \frac{z(\mathbf{x}_0 + h\mathbf{v}) - z(\mathbf{x}_0)}{h}. \tag{B.6}$$

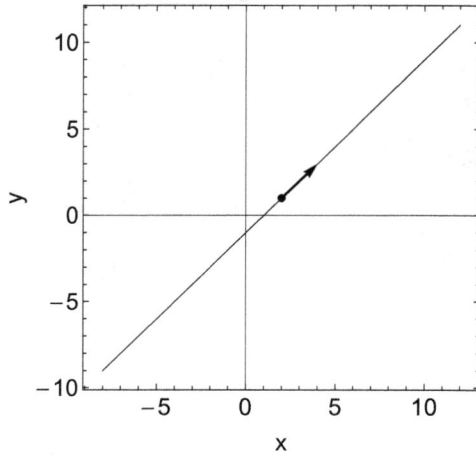

Fig. B.2.   A line function: The points in the graph shown in this figure are in the set produced using the expression $\mathbf{x_0} + \mathbf{v}t$, where $\mathbf{x_0} = (2, 1)$ and assuming $\mathbf{v} = (2, 2)$.

B.2   Gradients

Definition B.15 (Gradient). Let $z \colon \mathbb{R}^n \to \mathbb{R}$ be a function, and let $\mathbf{x_0} \in \mathbb{R}^n$. Then, the *gradient* of z at $\mathbf{x_0}$ is the vector in $\mathbb{R}^n$ given by

$$\nabla z(\mathbf{x_0}) = \begin{bmatrix} \frac{\partial z}{\partial x_1} \\ \vdots \\ \frac{\partial z}{\partial x_n} \end{bmatrix} \Bigg|_{\mathbf{x} = \mathbf{x_0}}. \tag{B.7}$$

Theorem B.16. *If $z \colon \mathbb{R}^n \to \mathbb{R}$ is a differentiable function, then all directional derivatives exist. Furthermore, the directional derivative of z at $\mathbf{x_0}$ in the direction of $\mathbf{v}$ is given by*

$$\nabla z(\mathbf{x_0}) \cdot \mathbf{v}. \tag{B.8}$$

Remark B.17. We now come to the two most important results about gradients: (i) they always point in the direction of steepest increase with respect to the level curves of a function; (ii) they are perpendicular (normal) to the level curves of the function used to compute them. These facts are exploited when we seek to maximize (or minimize) functions.

Theorem B.18. *Let $z: \mathbb{R}^n \to \mathbb{R}$ be a differentiable function, and let $\mathbf{x}_0 \in \mathbb{R}^n$. If $\nabla z(\mathbf{x}_0) \neq 0$, then $\nabla z(\mathbf{x}_0)$ points in the direction in which z is increasing fastest.*

Proof. Recall that $\nabla z(\mathbf{x}_0) \cdot \mathbf{n}$ is the directional derivative of z in the direction $\mathbf{n}$ at $\mathbf{x}_0$. Assume that $\mathbf{n}$ is a unit vector (i.e., $\|\mathbf{n}\| = 1$). We know that

$$\nabla z(\mathbf{x}_0) \cdot \mathbf{n} = \|\nabla z(\mathbf{x}_0)\| \cos \theta, \tag{B.9}$$

where θ is the angle between the vectors $\nabla z(\mathbf{x}_0)$ and $\mathbf{n}$. The function $\cos \theta$ is largest when $\theta = 0$, that is, when $\mathbf{n}$ and $\nabla z(\mathbf{x}_0)$ are parallel vectors. (If $\nabla z(\mathbf{x}_0) = 0$, then the directional derivative is zero in all directions.) □

Theorem B.19. *Let $z: \mathbb{R}^n \to \mathbb{R}$ be differentiable, and let $\mathbf{x}_0$ lie in the level set $L_c(z)$ defined by $z(\mathbf{x}) = c$ for fixed $c \in \mathbb{R}$. Then, $\nabla z(\mathbf{x}_0)$ is normal to the set $L_c(z)$ in the sense that if $\mathbf{v}$ is a tangent vector at $t = 0$ of a path $\mathbf{r}(t)$ contained entirely in $L_c(z)$ with $\mathbf{r}(0) = \mathbf{x}_0$, then $\nabla z(\mathbf{x}_0) \cdot \mathbf{v} = 0$.*

Proof. As stated, let $\mathbf{r}(t)$ be a curve in $L_c(z)$. Then, $\mathbf{r}: \mathbb{R} \to \mathbb{R}^n$ and $z[\mathbf{r}(t)] = c$ for all $t \in \mathbb{R}$. Let $\mathbf{v}$ be the tangent vector to $\mathbf{r}$ at $t = 0$; that is,

$$\frac{d\mathbf{r}(t)}{dt}\bigg|_{t=0} = \mathbf{v}. \tag{B.10}$$

Differentiating $z[\mathbf{r}(t)]$ with respect to t using the chain rule and evaluating at $t = 0$ yields

$$\frac{d}{dt}z[\mathbf{r}(t)]\bigg|_{t=0} = \nabla z[\mathbf{r}(0)] \cdot \mathbf{v} = \nabla z(\mathbf{x}_0) \cdot \mathbf{v} = 0 = \frac{dc}{dt}. \tag{B.11}$$

Thus, $\nabla z(\mathbf{x}_0)$ is perpendicular to $\mathbf{v}$ and therefore normal to the set $L_c(z)$, as required. □

Example B.20. We illustrate this theorem in Fig. B.3. The function is $z(x, y) = x^4 + y^2 + 2xy$ and $\mathbf{x}_0 = (1, 1)$. At this point, $\nabla z(\mathbf{x}_0) = \langle 6, 4 \rangle$. In the figure, we have scaled the vector to improve legibility.

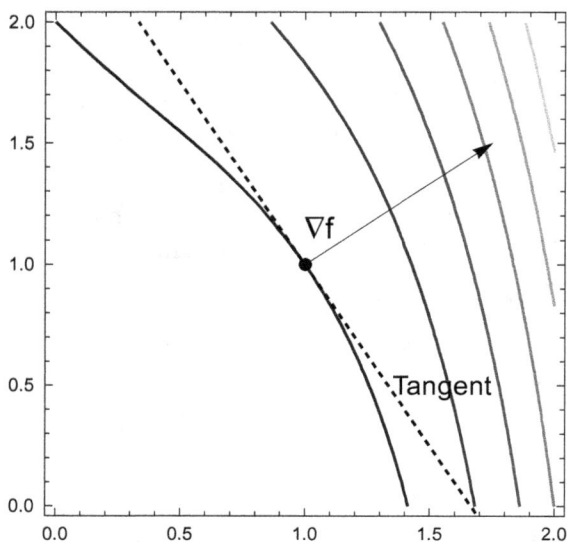

Fig. B.3.   A level curve plot with gradient vector. We've scaled the gradient vector to make the picture reasonably compact. Note that the gradient is perpendicular to the level set curve at the point $(1, 1)$, where the gradient was evaluated.

$$- \spadesuit \clubsuit \heartsuit \diamondsuit -$$

B.3   Exercises

B.1   Use the value of the cosine function and the fact that $\mathbf{x} \cdot \mathbf{y} = ||\mathbf{x}|| \, ||\mathbf{y}|| \cos \theta$ to prove the lemma. [Hint: For what values of θ is $\cos \theta > 0$?]

B.2   Prove Proposition B.14. [Hint: Use the definition of derivative for a univariate function, apply it to the definition of directional derivative, and evaluate $t = 0$.]

B.3   In this exercise, you will use elementary calculus (and a little bit of vector algebra) to show that the gradient of a simple function is perpendicular to its level sets:

(a) Plot the level sets of $z(x, y) = x^2 + y^2$. Draw the gradient at the point $(x, y) = (2, 0)$. Convince yourself that it is normal to the level set $x^2 + y^2 = 4$.

(b) Now, choose any level set $x^2 + y^2 = k$. Use implicit differentiation to find dy/dx. This is the slope of a tangent line to the circle $x^2 + y^2 = k$. Let (x_0, y_0) be a point on this circle.

(c) Find an expression for a vector parallel to the tangent line at (x_0, y_0). [Hint: You can use the slope you just found.]

(d) Compute the gradient of z at (x_0, y_0), and use it and the vector expression you just computed to show that two vectors are perpendicular. [Hint: Use the dot product.]

Introduction to Evolutionary Games Using the Replicator Equation

Appendix Goals: This appendix contains a brief introduction to evolutionary game theory using replicator dynamics. It is designed to be a self-contained introduction. The goals of the appendix are to introduce differential equations and systems, define fixed points of a system of ordinary differential equations, and illustrate the phase portrait of a system. We then discuss stability concepts, derive the replicator, and show some examples.

C.1   Differential Equations

Definition C.1. An *ordinary differential equation* (ODE) is an equation involving an *unknown* function of one variable, such as $y(x)$ or $x(t)$, and any number of its derivatives. We generally assume that the dependent variable is restricted to a known interval I, which could be $\mathbb{R}$.

Remark C.2 (Notation). Let $y(x)$ be a function of the independent variable x. Then, we can write the first derivative of y as

$$\frac{dy}{dx} = y'(x). \tag{C.1}$$

The nth derivative can be written as

$$\frac{d^n y}{dx^n} = y^{(n)}(x), \tag{C.2}$$

with the second derivative usually written $y''(x)$ and the third derivative written $y'''(x)$. The exception to this notation is when the independent variable is time (or time-like), in which case we may have a function $x(t)$. Then, the first derivative can (but doesn't have to be) written as

$$\frac{dx}{dt} = \dot{x}, \tag{C.3}$$

$$\frac{d^2 x}{dt^2} = \ddot{x}, \tag{C.4}$$

$$\vdots \tag{C.5}$$

$$\frac{d^n x}{dt^n} = x^{(n)}(t). \tag{C.6}$$

The "dot notation" is a holdover from Newton's fluxion notation, while the rest of the notation is due to Leibniz.

Remark C.3 (Notation Remark). For the remainder of this appendix, we will use the dot notation, as we will be taking derivatives with respect to time.

Remark C.4. In general, we can write an ODE as

$$F(x, \dot{x}, \ldots, x^{(n)}, t) = 0, \tag{C.7}$$

where F represents a function acting on the unknown function $x(t)$, its derivatives, and its independent variable t.

Definition C.5 (Order). Consider the ODE $F(x, \dot{x}, \ldots, x^{(n)}, t) = g(t)$. The *order* of the ODE is n, the degree of the highest derivative appearing in the equation.

Example C.6 (Exponential Growth/Decay). The following is a simple first-order ODE:

$$\dot{x} - \alpha x = 0, \tag{C.8}$$

where $\alpha \in \mathbb{R}$ is a constant. This differential equation models exponential growth (or decay). We study it in much greater detail shortly.

Definition C.7 (Initial Value Problem). An initial value problem (IVP) is a differential equation $F(x, \dot{x}, \ldots, x^{(n)}, t) = 0$ along with the conditions

$$x(t_0) = x_0,$$
$$\dot{x}(t_0) = x_1,$$
$$\vdots$$
$$x^{(n-1)}(t_0) = x_{n-1}.$$

Example C.8 (Exponential Growth/Decay). Consider the IVP

$$\dot{x} - \alpha x = 0,$$
$$x(0) = x_0 > 0.$$

We can write the ODE as

$$\frac{dx}{dt} = \alpha x.$$

Multiplying by dt, dividing by x, and integrating yields

$$\int \frac{dx}{x} = \int \alpha dt.$$

A little computation shows that

$$\log(x) = \alpha t + C,$$

where C is a constant of integration. You can put it on either side; however, by convention, it usually goes on the side with the independent variable. Since $x(0) = x_0$, it's easy to see that when $t = 0$, we have

$$\log(x_0) = C.$$

Thus, $\log(x) - \alpha t - \log(x_0) = 0$ implicitly solves the ODE. We can do better. Note that

$$\log(x) - \log(x_0) = \log\left(\frac{x}{x_0}\right) = \alpha t.$$

Taking the exponential yields

$$\frac{x}{x_0} = e^{\alpha t},$$

or

$$x(t) = x_0 e^{\alpha t}.$$

When $\alpha > 0$, the system is said to experience exponential growth. When $\alpha < 0$, the system is said to experience exponential decay. This model is frequently used to produce simple models of biological populations.

Remark C.9. Note that, if $x_0 > 0$ (as we assumed), then $x_0 e^{\alpha t} > 0$ for all t. If $x_0 < 0$, we need to use the (more correct) formula

$$\int \frac{dx}{x} = \log |x|.$$

The result is the same. If $x_0 \in \mathbb{R}$, we always obtain the solution $x(t) = x_0 e^{\alpha t}$ for the equation $\dot{x} = \alpha x$ with $x(0) = x_0$.

Definition C.10. A *system of ODEs* is a set of equations involving a set of unknown functions $y_1, \ldots, y_n$ and their derivatives, with each being a function of one independent variable.

Remark C.11. In this appendix, we focus on *autonomous* systems of first-order differential equations with the form

$$\begin{cases} \dot{x}_1 = f_1(x_1, \ldots, x_n), \\ \qquad \vdots \\ \dot{x}_n = f_n(x_1, \ldots, x_n). \end{cases} \tag{C.9}$$

Here, $x_1(t), \ldots, x_n(t)$ are n unknown functions with the independent variable t (time). We assume that for $i = 1, \ldots, n$, $f_i(x_1, \ldots, x_n)$ has derivatives of all orders. The system is *autonomous* because time does not appear explicitly on the right-hand sides of the equations. We may be given initial conditions for such a system in the form

$$x_1(0) = a_1, \ldots, x_n(0) = a_n,$$

where $a_1, \ldots, a_n$ are constants.

Example C.12. Consider the system of differential equations

$$\begin{cases} \dot{x} = -y, \\ \dot{y} = x, \end{cases} \tag{C.10}$$

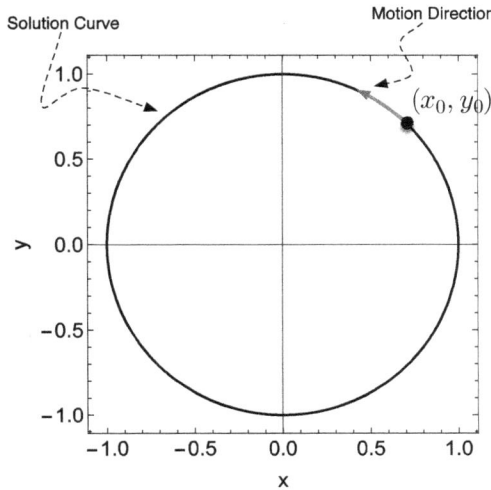

Fig. C.1. A solution curve for a specific initial condition (x_0, y_0) for Eq. (C.10).

with the initial condition $x(0) = x_0$ and $y(0) = y_0$. This system has the solution

$$x(t) = x_0 \cos(t) - y_0 \sin(t),$$

$$y(t) = x_0 \sin(t) + y_0 \cos(t),$$

which one can check by taking derivatives. For a given initial condition, (x_0, y_0), we can plot the parametric curve $\mathbf{r}(t) = \langle x(t), y(t) \rangle$, called a *solution curve*. The result is shown in Fig. C.1.

C.2 Fixed Points, Stability, and Phase Portraits

Remark C.13. The system in Eq. (C.9) can be written compactly as the vector equation

$$\dot{\mathbf{x}} = \mathbf{F}(\mathbf{x}), \tag{C.11}$$

where $\mathbf{F} : \mathbb{R}^n \to \mathbb{R}^n$ is a vector-valued function of a vector of inputs. That is,

$$\mathbf{F}(\mathbf{x}) = \begin{bmatrix} f_1(x_1, \ldots, x_n), \\ \vdots \\ f_n(x_1, \ldots, x_n). \end{bmatrix}.$$

Definition C.14 (Fixed Points). Consider the system of differential equations given by Eq. (C.11). A vector $\mathbf{x}^* \in \mathbb{R}^n$ is a *fixed* or *equilibrium* point of Eq. (C.11) if $\mathbf{F}(\mathbf{x}^*) = \mathbf{0}$.

Example C.15 (Lotka–Volterra Equations). Suppose x and y are the quantities of a prey species and a predator species, respectively. The prey species grows exponentially in the absence of any predators. The predators decay exponentially in the absence of any prey. When prey and predator are together, the prey are consumed by the predators and are removed at a rate proportional to the product of the numbers of predators and prey. Similarly, predators convert their consumed prey into new predators, which are added at a rate proportional to the product of the numbers of predator and prey.

The resulting system of differential equations describing this model is

$$\begin{cases} \dot{x} = \alpha x - \beta xy, \\ \dot{y} = \gamma xy - \delta y. \end{cases}$$

We can solve for the fixed points of this system by solving the system of nonlinear equations

$$\alpha x - \beta xy = 0,$$

$$\gamma xy - \delta y = 0.$$

Doing this yields two fixed points: $x^* = y^* = 0$ and

$$x^* = \frac{\delta}{\gamma},$$

$$y^* = \frac{\alpha}{\beta}.$$

This system of equations is called the Lotka–Volterra system.

Definition C.16 (Stable Fixed Point). Let

$$\dot{\mathbf{x}} = \mathbf{F}(\mathbf{x})$$

be an autonomous system of ODEs with the fixed point $\mathbf{x}^*$. Let $\mathbf{x}(t)$ be a solution curve with initial time t_0 and initial position $\mathbf{x}(t_0) = \mathbf{x}_0$. Then, $\mathbf{x}^*$ is *stable* if for all $\epsilon > 0$, there exists a $\delta > 0$ so that if $\|\mathbf{x}_0 - \mathbf{x}^*\| < \delta$, then for all $t > t_0$, $\|\mathbf{x}(t) - \mathbf{x}^*\| < \epsilon$.

Remark C.17. The stability of a fixed point sounds complex, but it's relatively simple conceptually. If a solution starts near a fixed point and stays near that fixed point over time, then the fixed point is stable. There is a stronger form of stability, called asymptotic stability, in which the trajectory approaches the fixed point in the long run.

Definition C.18. A fixed point, $\mathbf{x}^*$, is asymptotically stable if it is both stable and there is a $\delta_0 > 0$ so that if $\|\mathbf{x}_0 - \mathbf{x}^*\| < \delta_0$, then

$$\lim_{t \to \infty} x(t) = x^*.$$

Remark C.19. Fixed points can be further categorized. This is discussed at length in texts on nonlinear dynamics. Strogatz provides a thorough introduction [120]. It is possible to determine the stability of fixed points in many (but not all) settings using some basic calculus. For two-dimensional systems, a picture called a phase portrait can generally show whether a fixed point is stable, asymptotically stable, or neither (usually called unstable).

Definition C.20 (Phase Portrait). Consider Eq. (C.9), and suppose we have a general solution, $x_1(t; a_1), \ldots, x_n(t; a_n)$, parameterized by the initial values $a_1, \ldots, a_n$. (That is, suppose we have a specific orbit.) Then, a *phase portrait* is a geometric representation of the behavior of the differential equation system obtained by creating multiple parametric curves $(x_1(t; a_1), \ldots x_n(t; a_n)$ for various starting values $a_1, \ldots, a_n$, i.e., we are visualizing multiple solutions of the system.

Example C.21. Take a specific instance of the Lotka–Volterra dynamics:

$$\begin{cases} \dot{x} = x - xy, \\ \dot{y} = xy - y. \end{cases}$$

This system has two fixed points: $x^* = y^* = 0$ and $x^* = y^* = 1$. A phase portrait can help determine the stability of the fixed points. This is shown in Fig. C.2. The figure shows that the fixed point $x^* = y^* = 1$ is stable, but not asymptotically stable. This is because the solutions orbit around the fixed point, never getting too far away, just like in Fig. C.1. The fixed point $x^* = y^* = 0$ is not stable since

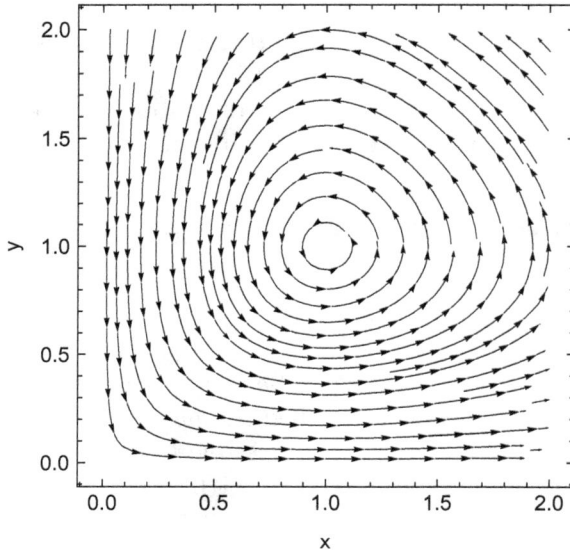

Fig. C.2. A phase portrait for specific Lotka–Volterra dynamics shows that one fixed point is stable, but not asymptotically stable. The other fixed point is not stable.

the solution curves can extend arbitrarily far away from the fixed point, depending on the initial condition.

C.3   The Replicator Equation

Remark C.22. We now derive a specific class of differential equations, called the replicator equations, using a game matrix and a biological interpretation of game play.

Derivation C.23. Suppose there are n species of organisms that interact in a pairwise fashion by playing a symmetric[1] game (such as prisoner's dilemma or rock-paper-scissors). Let $\mathbf{A} \in \mathbb{R}^{n \times n}$ be the payoff matrix for this game so that when a member of species i interacts with species j, it receives payoff, $\mathbf{e}_i^T \mathbf{A} \mathbf{e_j}$.

---

[1]Although this assumption is not needed, it makes the derivation sensible.

Suppose there are $X_i \geq 0$ members of the species i in the population, with $i \in \{1, \ldots, n\}$. For simplicity, we allow fractional population counts since we are building a notional model. Let

$$M = \sum_{i=1}^{n} X_i, \tag{C.12}$$

be the total population. Then,

$$x_i = \frac{X_i}{M}$$

is the proportion of the population composed of the species i. The vector $\mathbf{x} = \langle x_1, \ldots, x_n \rangle$ is the proportion vector. Since we associate species with strategies in the game, $\mathbf{x}$ acts like a mixed strategy.

Now, suppose that the fitness of the species i is its expected payoff in the game. That is,

$$f_i(\mathbf{x}) = \mathbf{e}_i^T \mathbf{A} \mathbf{x}.$$

In a game-theoretic sense, this is the expected payoff to a player playing a pure strategy, i, against a mixed strategy, $\mathbf{x}$. Now, suppose that the population of each species grows (or decays) exponentially with parameter f_i. That is,

$$\dot{X}_i = f_i X_i = X_i \left(\mathbf{e}_i^T \mathbf{A} \mathbf{x} \right). \tag{C.13}$$

This is a reasonable assumption in the sense that populations with positive fitness will grow, while those with negative fitness will shrink.

Raw population counts are unwieldy. We seek to compute $\dot{x}_i$, the rate of change of the proportion of species i in the whole population. That is, we want

$$\dot{x}_i = \frac{d}{dt} \left(\frac{X_i}{M} \right).$$

Both X_i and M are time-varying. Apply the quotient rule to find that

$$\dot{x}_i = \frac{\dot{X}_i M - X_i \dot{M}}{M^2} = \frac{\dot{X}_i}{M} - \frac{X_i}{M} \frac{\dot{M}}{M}.$$

Using the definition of $\dot{X}_i$ in Eq. (C.13), we have

$$\frac{\dot{X}_i}{M} = \frac{X_i}{M} \left(\mathbf{e}_i^T \mathbf{A} \mathbf{x} \right) = x_i \left(\mathbf{e}_i^T \mathbf{A} \mathbf{x} \right).$$

Now, use Eqs. (C.12) and (C.13) together to obtain

$$\frac{X_i}{M}\frac{\dot{M}}{M} = x_i\left(\frac{1}{M}\sum_{i=1}^{n}\dot{X}_i\right) = x_i\left(\frac{\sum_{i=1}^{n}X_i\left(\mathbf{e}_i^T\mathbf{A}\mathbf{x}\right)}{M}\right)$$

$$= x_i\left(\sum_{i=1}^{n}x_i\left(\mathbf{e}_i^T\mathbf{A}\mathbf{x}\right)\right) = x_i\left(\mathbf{x}^T\mathbf{A}\mathbf{x}\right).$$

Combining these together yields the expression

$$\dot{x}_i = x_i\left(\mathbf{e}_i^T\mathbf{A}\mathbf{x}\right) - x_i\left(\mathbf{x}^T\mathbf{A}\mathbf{x}\right) = x_i\left(\mathbf{e}_i^T\mathbf{A}\mathbf{x} - \mathbf{x}^T\mathbf{A}\mathbf{x}\right). \qquad \text{(C.14)}$$

This system of differential equations (for $i \in \{1,\ldots,n\}$) is called *replicator dynamics* or the *replicator equation*.

Remark C.24. In general, if $f_i(\mathbf{x})$ is a fitness function for a population in terms of population proportions and

$$\bar{f}(\mathbf{x}) = \sum_{i=1}^{n} x_i f_i(\mathbf{x})$$

is the mean fitness, then the replicator can be written as

$$\dot{x}_i = x_i\left(f_i(\mathbf{x}) - \bar{f}(\mathbf{x})\right).$$

In this way, the replicator can be generalized from fitness defined in terms of symmetric two-player games.

Remark C.25. A little work shows that as long as the initial conditions of the replicator dynamics are in Δ_n, i.e., $\mathbf{x}(0) = \mathbf{x}_0 \in \Delta_n$, then the dynamics remain within Δ_n. This requires showing that

$$\sum_{i=1}^{n}\dot{x}_i = 0.$$

Therefore, as time evolves, if $\mathbf{x}_0 \in \Delta_n$, then

$$\sum_{i=1}^{n}x_i(t) = 1,$$

for all time.

Remark C.26. We call the branch of game theory where we have populations dynamically changing (evolving) as a result of game play *evolutionary game theory*. There are several additional results in this area. However, there is a nice result relating Nash equilibria in symmetric games with fixed points in the replicator dynamics. We state but do not prove the *folk theorem of evolutionary games*, found in Ref. [121]. But first, we require a definition.

Definition C.27 (Interior of Δ_n). A point $\mathbf{x} = \langle x_1, \ldots, x_n \rangle \in \Delta_n$ is in the interior of Δ_n, written $\mathbf{x} \in \text{int}(\Delta_n)$ if $x_i > 0$ for all $i \in \{1, \ldots, n\}$.

Theorem C.28 (Folk Theorem of Evolutionary Games). *Let $\mathbf{A}$ be a game matrix for the row player in a two-player symmetric game. Then, the following hold:*

(1) *If $\mathbf{x}^*$ is a Nash equilibrium for the game, then $\mathbf{x}^*$ is a fixed point of the replicator dynamics defined by $\mathbf{A}$.*
(2) *If $\mathbf{x}^*$ is a strict Nash equilibrium, then $\mathbf{x}^*$ is an asymptotically stable fixed point of the replicator dynamics defined by $\mathbf{A}$.*
(3) *If $\mathbf{x}^*$ is a fixed point of the replicator dynamics in the interior of Δ_n and it is the limit as $t \to \infty$ of a solution curve lying entirely in the interior of Δ_n, then $\mathbf{x}^*$ is a Nash equilibrium.*
(4) *If $\mathbf{x}^*$ is a stable equilibrium point of the replicator dynamics, then it is a Nash equilibrium.*

Example C.29 (Prisoner's Dilemma). Consider the game matrix we have used for prisoner's dilemma:

$$\mathbf{A} = \begin{bmatrix} -1 & -10 \\ 0 & -5 \end{bmatrix},$$

and let $\mathbf{x} = \langle x_1, x_2 \rangle$. The replicator dynamics are the system of equations

$$\dot{x}_1 = x_1 \left(\underbrace{-x_1 - 10x_2}_{\mathbf{e}_1^T \mathbf{A} \mathbf{x}} - \underbrace{\left[(-10x_1 - 5x_2) x_2 - x_1^2 \right]}_{\mathbf{x}^T \mathbf{A} \mathbf{x}} \right),$$

$$\dot{x}_2 = x_2 \left(\underbrace{-5x_2}_{\mathbf{e}_2^T \mathbf{A} \mathbf{x}} - \underbrace{\left[(-10x_1 - 5x_2) x_2 - x_1^2 \right]}_{\mathbf{x}^T \mathbf{A} \mathbf{x}} \right).$$

If you carefully add these two expressions together and substitute in the fact that $x_2 = 1 - x_1$, you will see that they do sum to zero. One can also verify that the Nash equilibrium $\mathbf{e}_2$ is a fixed point of the dynamics. We note that not every fixed point needs to be a Nash equilibrium. For example, $\mathbf{x}^* = \mathbf{e}_1$ (the "cooperate" strategy) is also a fixed point; however, it is not a Nash equilibrium. This is due to the fact that if there are no players playing "defect," then that population will never grow.

From Theorem C.28, we expect $\mathbf{x}^* = \mathbf{e}_2$ to be an asymptotically stable fixed point. We can illustrate this by plotting some solution curves for $x_2(t)$. This is shown in Fig. C.3. From the figure, we see that unless $\mathbf{x}_0 = \mathbf{e}_1$, the solutions all approach $x_2 = 1$ (and hence $x_1 = 0$) as $t \to \infty$. That is, $\mathbf{e}_2$ is an asymptotically stable fixed point.

Example C.30 (Rock-Paper-Scissors). The dynamics get more interesting with three-strategy games. Consider the payoff matrix for rock-paper-scissors:

$$\mathbf{A} = \begin{bmatrix} 0 & -1 & 1 \\ 1 & 0 & -1 \\ -1 & 1 & 0 \end{bmatrix}.$$

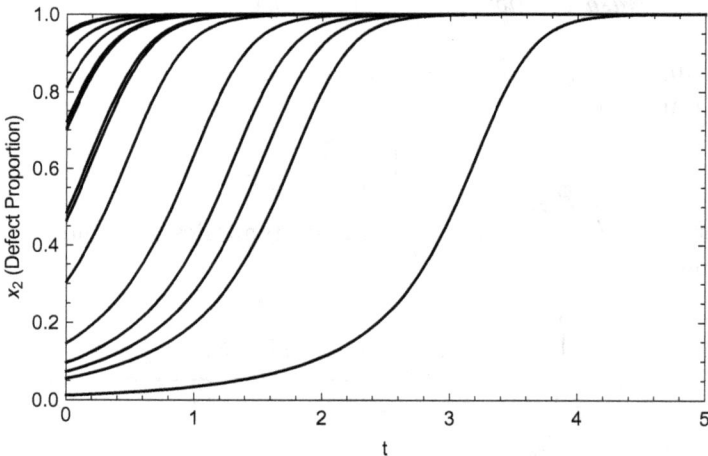

Fig. C.3. From any starting point, except $x_1 = 1$ and $x_2 = 0$, the solution to the replicator dynamics using the prisoner's dilemma matrix converges to $x_2 = 1$.

This matrix is interesting because

$$\mathbf{x}^T \mathbf{A} \mathbf{x} = 0. \tag{C.15}$$

Here, $\mathbf{x} = \langle x_1, x_2, x_3 \rangle$, with x_1, x_2, and x_3 being the proportions of the population playing rock, paper, and scissors, respectively. The resulting replicator dynamics become

$$\dot{x}_1 = x_1(x_3 - x_2),$$
$$\dot{x}_2 = x_2(x_1 - x_3),$$
$$\dot{x}_3 = x_3(x2 - x_1).$$

We know that $\mathbf{x}^* = \langle \frac{1}{3}, \frac{1}{3}, \frac{1}{3} \rangle$ is a fixed point since it is a Nash equilibrium. We can construct a plot to determine its stability. Recall that Δ_3 is a triangle in $\mathbb{R}^3$ (see Fig. 5.4). We can construct the trajectories of the dynamics and show them on this triangle, as they would exist in $\mathbb{R}^3$, but flattened onto the page. This is shown in Fig. C.4. This figure shows that the Nash equilibrium is a stable but not asymptotically stable fixed point. This figure is sensible from a biological perspective. If there are many players playing rock but

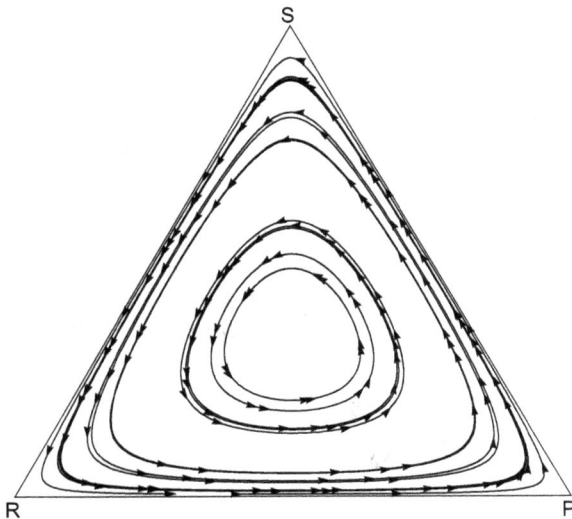

Fig. C.4. Example solution curves for rock-paper-scissors showing the cyclic nature of the game under the replicator dynamics.

only a few playing paper and scissors, then, ultimately, the paper players will "consume" the rock players, making more paper players, who will then be "consumed" by the scissors players, who are then "consumed" by the rock players, leading to the cyclic behavior.

Remark C.31. Intriguingly, we can modify the stability of the fixed point in the rock-paper-scissors game by adding a winner bias so that

$$\mathbf{A} = \begin{bmatrix} 0 & -1 & 1+a \\ 1+a & 0 & -1 \\ -1 & 1+a & 0 \end{bmatrix}. \tag{C.16}$$

The resulting theorem is a special case of a more general theorem of games like rock-paper-scissors that was proved by Zeeman [122], who characterized the behavior of the replicator dynamics for all three-strategy games. His result is summarized by Hofbauer and Sigmund in Ref. [121].

Theorem C.32. *Consider the replicator dynamics generated by the payoff matrix in Eq. (C.16). If $a > 0$, then the Nash equilibrium fixed point $\mathbf{x}^* = \left\langle \frac{1}{3}, \frac{1}{3}, \frac{1}{3} \right\rangle$ is asymptotically stable. If $a < 0$, then this fixed point is unstable. If $a = 0$, then the fixed point is stable but not asymptotically stable.*

Example C.33. We illustrate the previous theorem for the cases when $a = \frac{1}{10}$ and $a = -\frac{1}{10}$ in Fig. C.5. As stated in the theorem, when $a > 0$, the Nash equilibrium of the game is stable, and when $a < 0$, the Nash equilibrium of the game is unstable.

Remark C.34 (Bimatrix Games). We have assumed that the replicator dynamics are constructed from a symmetric game with Player 1 matrix $\mathbf{A}$, thereby using the fact that $\mathbf{B} = \mathbf{A}^T$. In the case when we have a non-symmetric game, we can still define a system of replicator equations, as discussed by Hofbauer in Ref. [123].

Suppose $\mathbf{A}, \mathbf{B} \in \mathbb{R}^{m \times n}$, and let $\mathbf{x} = \langle x_1, \ldots, x_m \rangle$ and $\mathbf{y} \in \langle y_1, \ldots, y_n \rangle$. The bimatrix replicator dynamics are

$$\begin{cases} \dot{x}_i = x_i \left(\mathbf{e}_i^T \mathbf{A} \mathbf{y} - \mathbf{x}^T \mathbf{A} \mathbf{y} \right) & i \in \{1, \ldots, m\}, \\ \dot{y}_j = y_j \left(\mathbf{x}^T \mathbf{B} \mathbf{e}_j - \mathbf{x}^T \mathbf{B} \mathbf{y} \right) & j \in \{1, \ldots, n\}. \end{cases}$$

Interestingly, if you use these dynamics in the symmetric game case, you sometimes get different behaviors than when using the ordinary

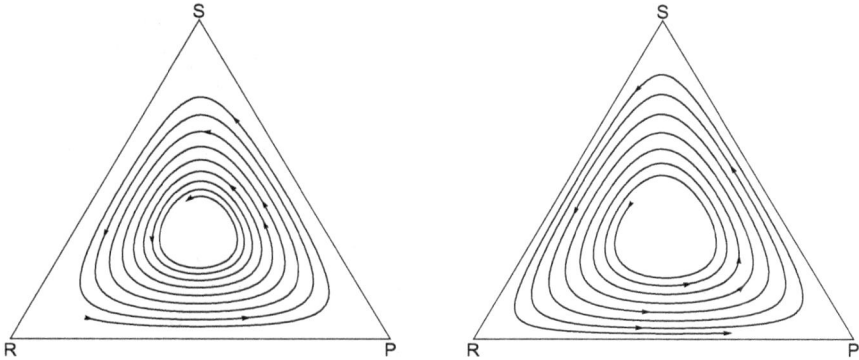

Fig. C.5. (Left) A sample trajectory from the biased rock-paper-scissors game, Eq. (C.16), when $a = \frac{1}{10}$. Here, the Nash equilibrium of the game is asymptotically stable. (Right) A sample trajectory from the biased rock-paper-scissors game, Eq. (C.16), when $a = -\frac{1}{10}$. Note that the Nash equilibrium of the game is unstable.

replicator dynamic with a single matrix (see Exercise C.2). This illustrates the importance of modeling choices when applying mathematics versus studying it in an abstract setting.

C.4 Appendix Notes

Evolutionary game theory originated with the work of Price and Smith [124]. Since this early work, the field has grown substantially. This appendix offers a mere glimpse into a much broader area of work that is summarized in books by Hofbauer and Sigmund [125], Weibull [126], and Friedman and Sinervo [127]. It is worth noting that most discussions on evolutionary games begin with the concept of evolutionary stability and evolutionarily stable strategies (ESS), which can be considered a refinement of the concept of Nash equilibrium. This is similar to the way that the Folk theorem (Theorem C.28) relates the stability of fixed points of the replicator equation to the Nash equilibria of the corresponding game. Evolutionary stability can be a subtle concept, which is why this brief introduction focuses on replicator dynamics, which are a bit more concrete and produce nice visualizations.

There are several variations of the replicator dynamics, some with more exotic properties. Examples include the replicator–mutator equation [128], which incorporates mutation into the replicator dynamics, allowing species to spontaneously emerge. Hofbauer and Sigmund [121] provide a more comprehensive list of variations. Several authors have modified the replicator equation to include spatial dynamics, thus creating partial differential equations. See Vickers' work for an example [129, 130]. Other authors have added higher-order interactions to model N-player games, including Gokhale and Traulsen [131] and Griffin and Wu [132]. Simple replicators can exhibit various forms of chaotic behavior (see, e.g., Refs. [133, 134]), making them interesting dynamical systems to study in isolation. Among other applications, replicator equations feature prominently in theoretical ecology, where they are used to study the qualitative properties of ecosystems [135, 136].

$$- \spadesuit \clubsuit \heartsuit \diamondsuit -$$

C.5 Exercises

C.1 Find the replicator dynamics for the Chicken game and compute its fixed points. Try to determine their stability by using a differential equation solver, like the one in Mathematica^{TM}.

C.2 Find the replicator dynamics for the Chicken game using the bimatrix formulation and compute its fixed points. Try to determine their stability by using a differential equation solver, like the one in Mathematica^{TM}.

C.3 Prove the statement in Remark C.25.

C.4 Explain why Theorems C.32 and C.28 do not contradict one another.

References

[1] R. D. Luce and H. Raiffa, *Games and Decisions: Introduction and Critical Survey*. Dover Press (1989).

[2] P. Morris, *Introduction to Game Theory*. Springer (1994).

[3] N. Nisan, T. Roughgarden, E. Tardos and V. V. Vazirani (eds.), *Algorithmic Game Theory*. Cambridge University Press (2007).

[4] J. González-Díaz, I. García-Jurado and M. G. Fiestras-Janeiro, *An Introductory Course on Mathematical Game Theory and Applications*, Vol. 238. American Mathematical Society (2023).

[5] J. von Neumann and O. Morgenstern, *The Theory of Games and Economic Behavior*, 60th edn. Princeton University Press (2004).

[6] J. Scarne, *Scarne's New Complete Guide to Gambling*. Simon & Schuster (1986).

[7] J. Archer, *The Archer Method of Winning at 21*. Henry Regnery Co. (1973).

[8] U. Today, Remembering 'let's make a deal' host monty hall, https://www.usatoday.com/picture-gallery/life/tv/2017/09/30/remembering-lets-make-a-deal-host-monty-hall/106172564/ (2017).

[9] C. B. Boyer and U. C. Merzbach, *A History of Mathematics*. John Wiley & Sons (2011).

[10] A. E. Moyer, Liber de ludo aleae, *Renaissance Quarterly* **60**, 4, pp. 1419–1420 (2007).

[11] K. Delvin, *The Unfinished Game: Pascal, Fermat, and the Seventeenth-Century Letter that Made the World Modern*. Basic Books (2008).

[12] O. Ore, Pascal and the invention of probability theory, *The American Mathematical Monthly* **67**, 5, pp. 409–419 (1960).

[13] L. J. Daston, Probabilistic expectation and rationality in classical probability theory, *Historia Mathematica* **7**, 3, pp. 234–260 (1980).

[14] T. Bayes, LII. An essay towards solving a problem in the doctrine of chances. By the late Rev. Mr. Bayes, F. R. S. communicated by Mr. Price, in a letter to John Canton, A. M. F. R. S, *Philosophical Transactions of the Royal Society of London*, 53, pp. 370–418 (1763).

[15] C. F. Gauß, *Theoria Motus Corporvm Coelestivm In Sectionibvs Conicis Solem Ambientivm*. Perthes et Besser (1809).

[16] N. L. Johnson, S. Kotz and N. Balakrishnan, *Continuous Univariate Distributions, Volume 1*, Vol. 289. John Wiley & Sons (1995).

[17] H. S. Bear, *A Primer of Lebesgue Integration*. Academic Press (2002).

[18] E. O. Thorp, *Beat the Dealer: A Winning Strategy for the Game of Twenty-one*, Vol. 310. Vintage (1966).

[19] J. L. Kelly, A new interpretation of information rate, *The Bell System Technical Journal* **35**, 4, pp. 917–926 (1956).

[20] B. Mezrich, *21: Bringing Down the House-Movie Tie-In: The Inside Story of Six MIT Students Who Took Vegas for Millions*. Simon & Schuster (2008).

[21] S. Selvin, Monty Hall problem, *American Statistician* **29**, 3, pp. 134–134 (1975).

[22] J. P. Morgan, N. R. Chaganty, R. C. Dahiya and M. J. Doviak, Let's make a deal: The player's dilemma, *The American Statistician* **45**, 4, pp. 284–287 (1991).

[23] R. G. Seymann, [let's make a deal: The player's dilemma]: Comment, *The American Statistician* **45**, 4, pp. 287–288 (1991).

[24] E. Barbeau, Fallacies, flaws, and flimflam, *College Mathematics Journal* **32**, 2, pp. 149–154 (1993).

[25] S. Lucas, J. Rosenhouse and A. Schepler, The Monty Hall problem, reconsidered, *Mathematics Magazine* **82**, 5, pp. 332–342 (2009).

[26] A. P. Flitney and D. Abbott, Quantum version of the monty hall problem, *Physical Review A* **65**, 6, p. 062318 (2002).

[27] J. Rosenhouse, *The Monty Hall Problem: The Remarkable Story of Math's most Contentious Brain Teaser*. Oxford University Press (2009).

[28] J. Tierney, Behind Monty Hall's doors: Puzzle, debate and answer, *New York Times* **21**, p. 1 (1991).

[29] E. Aslanian, 'the price is right' hits 9,000 episodes — the game show by the numbers, *TV Insider* (2019).

[30] G. Loomes, C. Starmer and R. Sugden, Observing violations of transitivity by experimental methods, *Econometrica: Journal of the Econometric Society*, **59**, 2, pp. 425–439 (1991).

[31] C. Blair Jr, Passing of a great mind, *Life* **25**, p. 96 (1957).

[32] F. Dyson, A meeting with Enrico Fermi, *Nature* **427**, 6972, pp. 297–297 (2004), https://doi.org/10.1038/427297a.

[33] J. Mayer, K. Khairy and J. Howard, Drawing an elephant with four complex parameters, *American Journal of Physics* **78**, 6, pp. 648–649 (2010).

[34] D. Bernoulli, *Specimen Theoriae Novae de Mensura Sortis: Translated into German and English*, Letter to Pierre Raymond de Montmort (1713).

[35] P. C. Fishburn, Utility theory, *Management Science* **14**, 5, pp. 335–378 (1968).

[36] E. Karni and D. Schmeidler, Utility theory with uncertainty, *Handbook of Mathematical Economics* **4**, pp. 1763–1831 (1991).

[37] N. G. Mankiw, *Principles of Microeconomic Theory*, 9th edn. Cenage (2021).

[38] R. B. Myerson, *Game Theory: Analysis of Conflict*. Harvard University Press (2001).

[39] C. H. Griffin, *Applied Graph Theory: An Introduction with Graph Optimization and Algebraic Graph Theory*. World Scientific (2023).

[40] S. J. Brams, *Game Theory and Politics*. Dover Press (2004).

[41] M. C. Fu, AlphaGo and Monte Carlo tree search: The simulation optimization perspective, in *2016 Winter Simulation Conference (WSC)*. IEEE, pp. 659–670 (2016).

[42] R. Bellman, Dynamic programming, *Science* **153**, 3731, pp. 34–37 (1966).

[43] S. B. Kotsiantis, Decision trees: A recent overview, *Artificial Intelligence Review* **39**, pp. 261–283 (2013).

[44] S. Russell and P. Norvig, *Artificial Intelligence: A Modern Approach*. Pearson (2016).

[45] D. E. Knuth and R. W. Moore, An analysis of alpha-beta pruning, *Artificial Intelligence* **6**, 4, pp. 293–326 (1975).

[46] F.-H. Hsu, *Behind Deep Blue: Building the Computer that Defeated the World Chess Champion*. Princeton University Press (2002).

[47] J. Schaeffer, N. Burch, Y. Bjornsson, A. Kishimoto, M. Muller, R. Lake, P. Lu and S. Sutphen, Checkers is solved, *Science* **317**, 5844, pp. 1518–1522 (2007).

[48] E. R. Berlekamp, J. H. Conway and R. K. Guy, *Winning Ways for Your Mathematical Plays*, Vol. 1. A. K. Peters (2001a).

[49] E. R. Berlekamp, J. H. Conway and R. K. Guy, *Winning Ways for Your Mathematical Plays*, Vol. 2. A. K. Peters (2001b).

[50] E. R. Berlekamp, J. H. Conway and R. K. Guy, *Winning Ways for Your Mathematical Plays*, Vol. 3. A. K. Peters (2001c).

[51] E. R. Berlekamp, J. H. Conway and R. K. Guy, *Winning Ways for Your Mathematical Plays*, Vol. 4. A. K. Peters (2001d).

[52] D. Silver, A. Huang, C. J. Maddison, A. Guez, L. Sifre, G. Van Den Driessche, J. Schrittwieser, I. Antonoglou, V. Panneershelvam,

M. Lanctot *et al.*, Mastering the game of go with deep neural networks and tree search, *Nature* **529**, 7587, pp. 484–489 (2016).

[53]   R. W. Rosenthal, Games of perfect information, predatory pricing and the chain-store paradox, *Journal of Economic theory* **25**, 1, pp. 92–100 (1981).

[54]   C. F. Camerer, *Behavioral Game Theory: Experiments in Strategic Interaction.* Princeton University Press (2011).

[55]   O. G. Haywood Jr, Military decision and game theory, *Journal of the Operations Research Society of America* **2**, 4, pp. 365–385 (1954).

[56]   I. Ravid, Military decision, game theory and intelligence: An anecdote, *Operations Research* **38**, 2, pp. 260–264 (1990).

[57]   R. Franck and F. Melese, A game theory view of military conflict in the Taiwan strait, *Defense & Security Analysis* **19**, 4, pp. 327–348 (2003).

[58]   W. N. Caballero, B. J. Lunday and R. F. Deckro, Leveraging behavioral game theory to inform military operations planning, *Military Operations Research* **25**, 1, pp. 5–22 (2020).

[59]   B. Burke, Game theory and run/pass balance, http://www.advancedfootballanalytics.com/2008/06/game-theory-and-runpass-balance.html (2008).

[60]   J. J. Sylvester, XLVII Additions to the articles in the September number of this journal, "on a new class of theorems," and on Pascal's Theorem, *The London, Edinburgh, and Dublin Philosophical Magazine and Journal of Science* **37**, 251, pp. 363–370 (1850).

[61]   J. Munkres, *Topology.* Prentice Hall (2000).

[62]   J. Nash, Non-cooperative games, *Annals of Mathematics*, **54**, 2, pp. 286–295 (1951).

[63]   J. Nash, C1 isometric imbeddings, *Annals of Mathematics*, **60**, 3, pp. 383–396 (1954).

[64]   J. Nash, The imbedding problem for Riemannian manifolds, *Annals of Mathematics* **63**, 1, pp. 20–63 (1956).

[65]   J. Moser, A rapidly convergent iteration method and non-linear partial differential equations-I, *Annali della Scuola Normale Superiore di Pisa-Scienze Fisiche e Matematiche* **20**, 2, pp. 265–315 (1966a).

[66]   J. Moser, A rapidly convergent iteration method and non-linear partial differential equations-II, *Annali della Scuola Normale Superiore di Pisa-Scienze Fisiche e Matematiche* **20**, 2, pp. 499–535 (1966b).

[67]   W. Rudin *et al.*, *Principles of Mathematical Analysis*, Vol. 3. McGraw-Hill New York (1976).

[68]   S. Nasar, *A Beautiful Mind.* Simon & Schuster (2011).

[69]   D. Bertsekas, *Nonlinear Programming*, 4th edn. Athena Scientific (2016).

[70] M. S. Bazaraa, H. D. Sherali and C. M. Shetty, *Nonlinear Programming: Theory and Algorithms*. John Wiley & Sons (2006).

[71] SciPy, SciPy v1.12.0 manual (scipy.optimize.linprog), https://docs. scipy.org/doc/scipy/reference/generated/scipy.optimize.linprog.html (2024).

[72] J. E. Marsden and A. Tromba, *Vector Calculus*, 5th edn. W. H. Freeman (2003).

[73] S. Lang, *Introduction to Linear Algebra*. Springer Science & Business Media (2012).

[74] H. W. Kuhn and A. W. Tucker, Nonlinear programming, in *Proceedings of the Second Berkeley Symposium on Mathematical Statistics and Probability*, California (1951).

[75] A. H. Foundation, William Karush, https://ahf.nuclearmuseum. org/ahf/profile/william-karush/ (2024).

[76] W. Karush, *Minima of Functions of Several Variables with Inequalities as Side Constraints* (1939).

[77] H. Kuhn and S. Nasar (eds.), *The Essential John Nash*. Princeton University Press (2002).

[78] M. S. Bazaraa, J. J. Jarvis and H. D. Sherali, *Linear Programming and Network Flows*. Wiley-Interscience (2004).

[79] S. P. Boyd and L. Vandenberghe, *Convex Optimization*. Cambridge University Press (2004).

[80] R. O. Duda, P. E. Hart and D. G. Stork, *Pattern Classification*. Wiley Interscience (2000).

[81] J. Nocedal and S. J. Wright, *Numerical Optimization*. Springer (1999).

[82] L. A. Wolsey and G. L. Nemhauser, *Integer and Combinatorial Optimization*. John Wiley & Sons (2014).

[83] G. Sierksma and Y. Zwols, *Linear and Integer Optimization: Theory and Practice*. CRC Press (2015).

[84] D. J. Albers and C. Reid, An interview with George B. Dantzig: The father of linear programming, *The College Mathematics Journal* **17**, 4, pp. 292–314 (1986).

[85] R. W. Cottle, George B. Dantzig: Operations research icon, *Operations Research* **53**, 6, pp. 892–898 (2005).

[86] R. W. Cottle, George B. Dantzig: A legendary life in mathematical programming, *Mathematical Programming* **105**, 1, pp. 1–8 (2006).

[87] G. B. Dantzig, Reminiscences about the origins of linear programming, in *Mathematical Programming the State of the Art*. Springer, pp. 78–86 (1983).

[88]   G. B. Dantzig, A proof of the equivalence of the programming problem and the game problem, *Activity Analysis of Production and Allocation* **13**, pp. 330–335 (1951).

[89]   T. Raghavan, Zero-sum two-person games, *Handbook of Game Theory with Economic Applications* **2**, pp. 735–768 (1994).

[90]   I. Adler, The equivalence of linear programs and zero-sum games, *International Journal of Game Theory* **42**, pp. 165–177 (2013).

[91]   L. G. Khachiyan, A polynomial algorithm in linear programming, in *Doklady Akademii Nauk*, Vol. 244. Russian Academy of Sciences, pp. 1093–1096 (1979).

[92]   N. Karmarkar, A new polynomial-time algorithm for linear programming, in *Proceedings of the Sixteenth Annual ACM Symposium on Theory of Computing*, pp. 302–311 (1984).

[93]   Gurobi Optimization, Gurobi, https://www.gurobi.com/academia/academic-program-and-licenses/ (2024).

[94]   IBM, IBM ILOG CPLEX Optimizer, https://www.ibm.com/products/ilog-cplex-optimization-studio/cplex-optimizer (2024).

[95]   O. L. Mangasarian and H. Stone, Two-person nonzero-sum games and quadratic programming, *Journal of Mathematical Analysis and Applications* **9**, pp. 348–355 (1964).

[96]   C. E. Lemke and J. T. Howson, Equilibrum points of bimatrix games, *Journal of the Society for Industrial and Applied Mathematics* **12**, 2, pp. 413–423 (1961).

[97]   R. Wilson, Computing equilibria of N-person games, *SIAM Journal on Applied Mathematics* **21**, 1, pp. 80–87 (1971).

[98]   R. W. Cottle, J.-S. Pang and R. E. Stone, *The Linear Complementarity Problem*. SIAM (2009).

[99]   T. Li and S. P. Sethi, A review of dynamic Stackelberg game models, *Discrete & Continuous Dynamical Systems-B* **22**, 1, p. 125 (2017).

[100]   S. P. Dirkse and M. C. Ferris, The path solver: A nommonotone stabilization scheme for mixed complementarity problems, *Optimization Methods and Software* **5**, 2, pp. 123–156 (1995).

[101]   M. Mavronicolas, V. Papadopoulou and P. Spirakis, Algorithmic game theory and applications, *Handbook of Applied Algorithms: Solving Scientific, Engineering and Practical Problems*, pp. 287–315 (2008).

[102]   G. Persiano, *Algorithmic Game Theory*. Springer (2011).

[103]   T. Roughgarden, Algorithmic game theory, *Communications of the ACM* **53**, 7, pp. 78–86 (2010).

[104]   N. Bitansky, O. Paneth and A. Rosen, On the cryptographic hardness of finding a Nash equilibrium, in *2015 IEEE 56th Annual Symposium on Foundations of Computer Science*. IEEE, pp. 1480–1498 (2015).

[105] C. Daskalakis, P. W. Goldberg and C. H. Papadimitriou, The complexity of computing a Nash equilibrium, *Communications of the ACM* **52**, 2, pp. 89–97 (2009).

[106] E. Hazan and R. Krauthgamer, How hard is it to approximate the best Nash equilibrium? *SIAM Journal on Computing* **40**, 1, pp. 79–91 (2011).

[107] R. Mehta, Constant rank two-player games are PPAD-hard, *SIAM Journal on Computing* **47**, 5, pp. 1858–1887 (2018).

[108] A. Rubinstein, Settling the complexity of computing approximate two-player Nash equilibria, *ACM SIGecom Exchanges* **15**, 2, pp. 45–49 (2017).

[109] J. L. Cohen, *Multiobjective Programming and Planning*. Dover (2003).

[110] J. Nash, Two-person cooperative games, *Econometrica: Journal of the Econometric Society*, pp. 128–140 (1953).

[111] A. Muthoo, *Bargaining Theory with Applications*. Cambridge University Press (1999).

[112] K. Avrachenkov, J. Elias, F. Martignon, G. Neglia and L. Petrosyan, Cooperative network design: A Nash bargaining solution approach, *Computer Networks* **83**, pp. 265–279 (2015).

[113] M. Bateni, M. Hajiaghayi, N. Immorlica and H. Mahini, The cooperative game theory foundations of network bargaining games, in *Automata, Languages and Programming: 37th International Colloquium, ICALP 2010, Bordeaux, France, July 6-10, 2010, Proceedings, Part I 37*. Springer, pp. 67–78 (2010).

[114] M. M. Hassan and A. Alamri, Virtual machine resource allocation for multimedia cloud: A Nash bargaining approach, *Procedia Computer Science* **34**, pp. 571–576 (2014).

[115] C. Liu, K. Li, Z. Tang and K. Li, Bargaining game-based scheduling for performance guarantees in cloud computing, *ACM Transactions on Modeling and Performance Evaluation of Computing Systems (TOMPECS)* **3**, 1, pp. 1–25 (2018).

[116] K. Miettinen, *Nonlinear Multiobjective Optimization*, Vol. 12. Springer Science & Business Media (1999).

[117] A. Navon, A. Shamsian, G. Chechik and E. Fetaya, Learning the Pareto front with hypernetworks, *arXiv preprint arXiv:2010.04104* (2020).

[118] D. B. Gillies, Solutions to general non-zero-sum games, *Contributions to the Theory of Games* **4**, 40, pp. 47–85 (1959).

[119] L. S. Shapley, *A Value for N-person Games*, RAND Number P-295 (1952).

[120]   S. H. Strogatz, *Nonlinear Dynamics and Chaos: With Applications to Physics, Biology, Chemistry, and Engineering.* CRC Press (2018).

[121]   J. Hofbauer and K. Sigmund, Evolutionary game dynamics, *Bulletin of the American Mathematical Society* **40**, 4, pp. 479–519 (2003).

[122]   E. C. Zeeman, Population dynamics from game theory, in *Global Theory of Dynamical Systems*, no. 819 in Springer Lecture Notes in Mathematics. Springer (1980).

[123]   J. Hofbauer, Evolutionary dynamics for bimatrix games: A Hamiltonian system? *Journal of Mathematical Biology* **34**, pp. 675–688 (1996).

[124]   J. M. Smith and G. R. Price, The logic of animal conflict, *Nature* **246**, 5427, pp. 15–18 (1973).

[125]   J. Hofbauer and K. Sigmund, *Evolutionary Games and Population Dynamics.* Cambridge University Press (1998).

[126]   J. W. Weibull, *Evolutionary Game Theory.* MIT Press (1997).

[127]   D. Friedman and B. Sinervo, *Evolutionary Games in Natural, Social, and Virtual Worlds.* Oxford University Press (2016).

[128]   M. A. Nowak, *Evolutionary Dynamics.* Harvard University Press (2006).

[129]   G. Vickers, Spatial patterns and ESS's, *Journal of Theoretical Biology* **140**, 1, pp. 129–135 (1989), http://www.sciencedirect.com/science/article/pii/S0022519389800335.

[130]   G. Vickers, Spatial patterns and travelling waves in population genetics, *Journal of Theoretical Biology* **150**, pp. 329–337 (1991).

[131]   C. S. Gokhale and A. Traulsen, Evolutionary games in the multiverse, *Proceedings of the National Academy of Sciences* **107**, 12, pp. 5500–5504 (2010).

[132]   C. Griffin and R. Wu, Higher-order dynamics in the replicator equation produce a limit cycle in rock-paper-scissors, *Europhysics Letters* **142**, 3, p. 33001 (2023).

[133]   B. Skyrms, Chaos in game dynamics, *Journal of Logic, Language and Information* **1**, 2, pp. 111–130 (1992).

[134]   J. Paik and C. Griffin, Completely integrable replicator dynamics associated to competitive networks, *Physical Review E* **107**, 5, p. L052202 (2023).

[135]   S. Allesina and J. M. Levine, A competitive network theory of species diversity, *Proceedings of the National Academy of Sciences* **108**, 14, pp. 5638–5642 (2011).

[136]   J. M. Levine, J. Bascompte, P. B. Adler and S. Allesina, Beyond pairwise mechanisms of species coexistence in complex communities, *Nature* **546**, 7656, pp. 56–64 (2017).

Index

$\alpha - \beta$ pruning, 81

A

affine function, 149
algorithmic game theory, 203
AlphaGo, 60

B

battle of Avranches, 103, 128
battle of the Bismark Sea, 59, 63, 73,
    76, 80, 82, 90
battle of the networks, 98
battle of the sexes (buddies), 209,
    225
Bayes' theorem, 25
Bernoulli, Johann, 47
best response function, 132
best response strategy, *see* strategy,
    best response
bimatrix replicator dynamics, *see*
    replicator dynamics (equation),
    bimatrix
blackjack, 19, 28
Blaise Pascal, *see* Pascal, Blaise
Bondareva–Shapley theorem,
    236
Brouwer fixed point theorem,
    125

C

card counting, 19, 28
    big player team, 28
characteristic (value) function, 230
chicken, 90–91, 93, 200
coalition, *see* cooperative game,
    coalition
coalition game, *see* cooperative game,
    coalition game
column vector, 244
competitive payoff region, *see* payoff
    region, competitive
complete information, 57
concave function, 148
conditional probability, *see*
    probability, conditional
conical combination, 146
constraint
    binding, 143
    equality, 141
    inequality, 141
convex combination, 146
convex function, 147
convex set, *see* set, convex
Conway, John, 81
cooperative game
    coalition game, 229–230, 232
    core, 234
    dominance, 233
    grand coalition, 229

imputation, 233
inessential game (zero-sum),
    232–233
cooperative payoff region, *see* payoff
    region, cooperative
coordinate game, *see* game,
    coordination
core, *see* cooperative game, core
craps, 15

D

Dantzig, George B., 183
de Mérée, Chevalier, 27
deal or no deal, 3, 6, 12
decision tree, *see* tree, decision
Deep Blue, 81
descendent, *see* tree, descendent
diagonal matrix, *see* matrix, diagonal
differential equation
    fixed point, 262
    initial value problem, 259
    order, 258
    phase portrait, 263
    system, 260
        autonomous, 260
differential equation, 257
directed graph, *see* graph, directed
directed path, *see* path, directed
directed tree, *see* tree, directed
directional derivative, 251–252
discrete probability distribution, *see*
    probability distribution
discrete probability space, *see*
    probability space
dominance (imputation), *see*
    cooperative game, dominance
dominated strategy, *see* strategy,
    dominated
dot product, 245, 249
dual feasibility, 152

E

edge
    move assignment, 57
    out, 54

elephant, 46
equilibrium, 76, 89, 94
    existence, 77, 125, 133
    Nash, 109, 115
    subgame perfect, 79
    zero-sum game, 98, 100, 120
evolutionary game theory, 267
expected utility theorem, 40, 42
expected value, 9
exponential growth (decay), 258–259

F

Fermat, Pierre de, 27
fixed point
    asymptotic stability, 263
    stability, 262
folk theorem of evolutionary games,
    267
football (North American), 88

G

game
    against the house, 3
    chance, 3, 65
    complete information, 57
    complexity, 60
    constant sum, 88
    coordination, 210
    general sum (KKT conditions), 88,
        191
    incomplete information, 61–62
    normal form, 87, 92
    strategic form, 89
    symmetric, 92
    value, 102
    zero-sum, 88
    zero-sum (linear program), *see*
        linear programming problem,
        zero-sum game
    zero-sum (matrix), 92
game tree, *see* tree, game
Gauss, Carl, 27
general-sum game, *see* game, constant
    sum, *see* game, general sum
generalized convexity, 157

Gillies, Donald, 239
global maximum, 141
gradient, 252–253
graph, 51
  directed, 51
graph (function), 250

H

Hawk-Dove, *see* Chicken
height, *see* tree, height
Huygens, Christiaan, 27

I

identity matrix, *see* matrix, identity
imputation, *see* cooperative game,
  imputation
independence, *see* probability,
  independent events
indifference theorem, 117, 182
information set, *see* set, information
initial value problem, *see* differential
  equation, initial value problem
inner product, 245
intersection, *see* set, intersection
iterative dominance, 117

K

Karush–Kuhn–Tucker theorem, 149,
  151
Kasparov, Gary, 81
Kelly, J. L., 28
Kolmogorov, Andrey, 28
Kuhn, Harold W., 157

L

lagrange multipliers, 151
Laplace, Pierre-Simon, 27
leave, *see* vertex, terminal
Lemke–Howson algorithm, 202
level set, *see* set, level, 253
line (function), 251
linear combination, 146
linear programming problem, 160
  computer solution, 172
  dual problem, 181, 235
  infinite solutions, 166

standard form, 173
zero-sum game, 168, 175
local maximum, 141
Lotka–Volterra equations, 262
lottery, 33–34
  compound, 36

M

Markov, Andrey, 28
matrix, 243
  addition, 244
  diagonal, 246
  identity, 246
  multiplication, 245
  scalar multiplication, 244
  symmetric, 247
  transpose, 245
minimax theorem, 121, 181
mixed strategy, *see* strategy, mixed
mixed-strategy space, 108
Monty Hall problem, 22, 29
Morgenstern, Oskar, 46
move assignment, *see* edge, move
  assignment
mutual assured destruction, 95

N

Nasar, Silvia, 134
Nash bargaining theorem, 220, 222
  axioms, 217
Nash equilibrium, *see* equilibrium,
  Nash
Nash, John F., 134
Nick the Greek, 11
norm, *see* vector, norm

O

one vector, *see* vector, one
optimization problem, 141
  general sum game, 190
  multi-criteria, 213, 227
    cooperative game, 216
ordinary differential equation, 257
out edge, *see* edge, out
outcome, *see* probability, outcome

P

Pareto frontier, 216, 226
Pareto optimality, 215
partition, 61
Pascal, Blaise, 27
path, 69
    directed, 52
payoff function, 57, 73
    expected, 73
        cooperative, 208
        mixed strategy, 108
payoff region
    competitive, 209
    cooperative, 209
perfect information strategy, *see*
    strategy, perfect inforamtion
phase portrait, *see* differential
    equation, phase portrait
Pierre de Fermat, *see* de Fermat,
    Pierre
player
    mixed-strategy space, 107
    player 0, 65
player vertex assignment, *see* vertex,
    player assignment
poker, 72
    coin, 66, 68, 72, 75
power set, *see* set, power set
preference, 35
    transitivity, 35
prisoner's dilemma, 111, 114,
    116, 267
probability, 3
    conditional, 15, 17
    event, 4
        mutually exclusive events, 5
    history, 27
    independent events, 18
    outcome, 4
    sample space, 4
probability distribution, 5
probability space, 6
pseudo-convex function, 157

Q

quadratic programming problem, 187
    computer solution, 189
    general sum game, 193
quasi-convex function, 157

R

random variable, 9
rationality, 59
replicator dynamics (equation), 264
    bimatrix, 270
rock-paper-scissors, 57, 106, 109, 119,
    268
Roth, Alvin, 239
roulette, 10
row vector, 244

S

saddle point, 102, 130
sample space, *see* probability, event,
    see probability, sample space
set
    bounded, 212
    closed, 212
    convex, 145–146, 211
    information, 62
    intersection, 5
    level, 250
    power set, 5
    stable, 234
    union, 5
set, convex, 148
Shapley, Lloyd, 239
Shapley values, 237
slack variable, 174
Snowdrift, *see* Chicken
St. Petersburg paradox, 47
stable fixed point, *see* fixed point,
    stability
stable set, *see* set, stable
strategy
    best response, 131

column dominance, 115
cooperative mixed, 208
dominated, 111
imperfect information, 63
mixed, 105
Nash, *see* equilibrium, Nash
perfect information, 58
pure, 108
row dominance, 114
strict dominance, 111
weak dominance, 111
strategy space, 73
strong duality theorem,
181
subtree, *see* tree, subtree
superadditivity, 231
surplus variable, 174
symmetric matrix, *see* matrix,
symmetric

T

tangent vector, *see* vector,
tangent
terminal vertex, *see* vertex, terminal
The Price is Right, 33–34
Thorp, Edward O., 28
transpose, *see* matrix, transpose
tree, 52
decision, 81
descendent, 54
directed, 52
game, 56–57, 62, 65, 67
height, 54
subtree, 55
Tucker, Albert W., 157

U

union, *see* set, union
unit simplex, 108
utility function
affine transformation, 46
linear, 45
utility theory, 47

V

vector
mixed strategy, 107
norm, 249
one, 247
pure strategy, 108
standard basis, 93, 246
tangent, 253
zero, 247
vertex
player assignment, 56
terminal, 54
von Neumann, John, 46, 183

W

weak dominance, *see* strategy, strict
dominance, *see* strategy, weak
dominance
Weierstrass' theorem, 219
Wilson's theorem, 202

Z

Zeeman's theorem, 270
Zermelo's theorem, 81
zero vector, *see* vector, zero
zero-sum game, *see* game, zero-sum

www.ingramcontent.com/pod-product-compliance
Lightning Source LLC
Chambersburg PA
CBHW050636190326
41458CB00008B/2290